Embryology, Genetics, and Ecology

– Windows into the Dynamic Nature of Living Organisms –

Embryology, Genetics, and Ecology

— Windows into the Dynamic Nature of Living Organisms —

Michael Holdrege

Waldorf
PUBLICATIONS
RESEARCH INSTITUTE FOR Waldorf EDUCATION

Printed with support from the Waldorf Curriculum Fund

Published by
Waldorf Publications at the
Research Institute for Waldorf Education
351 Fairview Avenue
Suite 625
Hudson, NY 12534

Title: *Embryology, Genetics, and Ecology*
Windows into the Dynamic Nature of Living Organisms
Author: Michael Holdrege
Drawings: Michael Holdrege
Proofreader: Mary Stowe
Layout: Ann Erwin
Cover image: Adobe Stock with permission

ISBN #978-1-943582-84-6

Table of Contents

Introduction

From the Whole to the Parts

In the 11th grade biology block we build upon the foundations laid in the 9th and 10th grades and carry them one step further. In the 9th grade we worked to firm up the student's ability to grasp straightforward linear causality. We studied those aspects of the human being that lent themselves to such an understanding. In the 10th grade we explored several organ systems, emphasizing their multifaceted functions and the complex processes by which various organs are interwoven with each other. This called on a form of dynamic, process-oriented thinking that had not been spoken to in the 9th grade. In the 11th grade we expand out in two directions from the organ systems studied in the 10th grade. On the one hand, we "move up a level" and work with several characteristics of organisms as a whole, and on the other, we "move down a level" to the "subunits" of organs, to the cellular dimension. Yet another level of organic activity—how organisms interact with each other within a larger ecological context—is be explored in the third part of this book

A central question that permeates this entire block is the relationship of the parts to the whole, of the whole to the parts. It is not only a fundamental question for all of biology, but also one that speaks to students at this age. As they look ahead to the world they will soon be moving out into, essential questions begin to awaken for them. They are wondering how issues they are facing and the decisions they need to make fit into the larger context of their entire biography, and how the ideals that are awakening in them could make a difference in the world they are growing up into.

They have moved beyond the quick reflexive judgments of their early teen years and are now able to ponder things from a larger perspective. In this course we seek to nurture and strengthen the capacity—and predisposition—to see not only attributes, but the context in which they appear; we are helping the students develop a sense for when an overriding whole is at work that is more than just an aggregate of its components.

The Drama of Scientific Discovery

This block provides another opportunity, as well. It gives us the chance to tell the amazing story of how central areas of modern biology have developed over the past 150 years. Much of what is studied in 11th grade biology provides excellent material for learning about the nature of science as a process (as a verb!) that leads scientists on a long and winding road of discoveries—often surprising and unexpected ones that are much more than a mere accumulation of facts upon which ironclad theories can be erected. There are big surprises and much drama in this unfolding that can be revealed if teachers free themselves from what Thomas Kuhn in his *Structure of Scientific Revolutions* calls "the textbook tradition." This tradition gives the impression, he says, that "one by one, in a process often compared to the addition of bricks to a building, scientists have added another fact, concept, law or theory to the body of information supplied in the contemporary scientific text. But that is not the way a science develops." (Kuhn, 1970, p. 140)

Kuhn devotes an entire chapter to describing how the textbook tradition has made it so difficult to see the scientific revolutions that have taken place in the past. He points to an ongoing change in the questions and answers pursued by scientists that spearhead the research in their fields of study. This is often overlooked because "the textbook tendency to make the development of science linear hides a process that lies at the heart of the most significant episodes of scientific development."

In a Waldorf science block, we are not bound by that tradition. We have the possibility of showing the students how the questions around developmental biology and genetics have been formulated and reformulated again and again. This comes about because new—and often unexpected—discoveries are sparked by the revision of research questions and the changes in perspective that they bring with them. The amazing mysteries revealed by the study of developing organisms will take us through the 20th century and up to the present moment, when the students will clearly see that modern biology is far from having reached the end of its road. In fact, several "central dogmas" of 20th century biology are now undergoing serious revision.[1]

As Evelyn Fox Keller, Professor of History and Philosophy of Science at MIT, already pointed out at the turn of the century:

1 In the volume *Everything Flows* (2018), edited by Daniel Nicholson and John Dupré and published by Oxford University Press, the reader can find an extensive and multifaceted discussion of why the *machine conception of the organism* (MCO) represents a dated perspective that does not do justice to living organisms.

> *For almost 50 years, we lulled ourselves into believing that, in discovering the molecular basis of genetic information, we had found the 'secret of life'.... And even though the message has yet to reach the popular press, to an increasingly large number of workers at the forefront of contemporary research, it seems evident that the primacy of the genes as the core explanatory concept of biological structure and function is more a feature of the 20th century than it will be of the 21st. What will take its place?... What in short, will the biology of the 21st century look like?* (Keller, 2002, pp. 7–9)

We are now well into the 21st century, and the discoveries and thinking that have brought a revolution to this field are there for our exploration as we move through this block! It is a very exciting story and one that is able to engage the students on many levels—since everyone enjoys a good drama!

Overview of What Is Ahead

Part I. Developmental Biology

Since a key goal of this block is to illuminate the relationship between the whole and the parts, I have found an effective way to begin is with a phase in the life of an organism where that relationship presents itself in a most dynamic and tangible way: during embryonic development. In contrast to the picture many people carry of an organism that is built up of an assembly of its primary parts, the cells, the students will learn that things are actually the other way around. The fertilized egg cell (zygote) is an organism that subdivides itself again and again as it moves through various developmental stages (morula, blastula, gastrula, etc.). It is the organism that differentiates itself, producing cells and bodily structures, it is not the cells that are combining to make the organism!

This lays the foundation for understanding the next big step in developmental biology: the birth of "experimental embryology." The experimental mindset that had grown so strong in other realms of science eventually takes hold of embryological studies, too. The emergence of two opposing schools of thought, the "mosaic" and "regulation" theories, raises very interesting questions that can only be resolved with the help of innovative experimentation. In this context, the students learn how hypotheses were formulated and experiments created to test them and how new instruments had to be fabricated, methods refined, and experimental results verified.

This all becomes very concrete when the students are introduced to a series of experiments performed by Hans Spemann and his team of graduate students at the University of Freiburg, which would lead to a Nobel Prize.[2] Going through these experiments the students see repeatedly how creative imagination played a pivotal role in their conception and implementation, and how new conceptual frameworks arise from the experimental findings. Above all, discoveries made through the Spemann experiments give the students a whole new sense for what the expression "organism" means. Through the concepts that are born of theses experimental findings—such as prospective potential, correspondent development, induction, etc.—they learn about the complex process known as "morphogenesis." It becomes clear that the generation of form and structure in the development of an organism requires the dynamic coordination of a wide range of sub-processes, all of which are interwoven spatially and temporally with each other. These discoveries culminate in what many consider to be "the single most important experiment in the history of embryological development (Shubin 2009, p. 106)."

Once a foundation is laid in this way, the microscopic dimension of cellular biology can be understood in the larger context of the organism as a whole. The centrality of controlled cell division and regulated cell death (apoptosis) becomes

2 Questions about the ethics of experimental research on animals lives strongly in today's youth. Therefore, it is helpful in this context to explain to them how consciousness around these questions has increased greatly in recent decades. It can be helpful for some classes to receive excerpts from a statement about such questions such as the following that can be found at the NIH's National Library of Medicine:

"Animal experimentation is widely used around the world for the identification of the root causes of various diseases in humans and animals and for exploring treatment options. Among the several animal species, rats, mice and purpose-bred birds comprise almost 90% of the animals that are used for research purposes. However, growing awareness of the sentience of animals and their experience of pain and suffering has led to strong opposition to animal research among many scientists and the general public. In addition, the usefulness of extrapolating animal data to humans has been questioned. This has led to Ethical Committees' adoption of the 'four Rs' principles (Reduction, Refinement, Replacement and Responsibility) as a guide when making decisions regarding animal experimentation. Due to the ethical issues surrounding the use of animals in experimentation, their use is declining in those research areas where alternative in vitro or in silico methods are available. However, so far it has not been possible to dispense with experimental animals completely and further research is needed to provide a road map to robust alternatives before their use can be fully discontinued.

"The first 'R,' Reduction, means that the experimental design is examined to ensure that researchers have reduced the number of experimental animals in a research project to the minimum required for reliable data. The second 'R,' Refinement, involves improvements in procedure that minimize the harmful effects of the proposed experiments on the animals involved, such as reducing pain, distress, and suffering in a manner that leads to a general improvement in animal welfare. The third 'R,' Replacement, refers to approaches that replace or avoid the use of experimental animals altogether. The fourth 'R,' Responsibility, refers to concerns around promoting animal welfare by improvements in experimental animals' social life, development of advanced scientific methods for objectively determining sentience, consciousness, experience of pain, and intelligence in the animal kingdom, as well as effective involvement in the professionalization of the public discussion on animal ethics." (Kiani A. et al., 2022)

more vivid, as does the importance of ongoing cytoskeleton remodeling and the convertibility of cell organelles based on the momentary needs of the organism. The pivotal role of cell differentiation is also a central topic in this context. The relationship of part to whole revealed by regeneration experiments is particularly eye-opening for the students. To everyone's amazement, some cells actually lose their differentiation and then re-differentiate as another cell type (transdifferentiation) so that they can fulfill a new function within the larger whole. The different way that cells develop "in vitro" and "in vivo" is also very enlightening in this regard.

Part II. Genetics

After immersing themselves for a time in embryological phenomena, the students are able to recognize quite quickly the very different kind of thinking that informs the second central topic of this block: the field of genetics as it unfolds from Mendel and Morgan, to the double helix and the Human Genome Project.

We now take the students on a journey that explores how the field of genetics arose and how it "rapidly developed into a separate discipline that had little contact with other branches of biology" (Morange 1998). What one finds on this journey provides the student with important insights into how:

- a scientific discipline can come into being and evolve over time;
- the kinds of questions that scientists ask—the mindset they bring to the task—greatly influences the nature of their findings and the conclusions they draw;
- what appear to be unshakeable findings at certain points in time can later be seen in a very different light—even turned upside down—as the science progresses and paradigms shift.

Starting with Mendel, the students learn how he carried very concrete questions about the nature of heredity and how he pursued them for decades with great tenacity and dedication. In the process, Mendel developed a unique approach to answering his questions: He focused on certain discrete traits rather than the entire organism, he assessed his findings quantitatively, and in order to explain those findings, he developed a thought model that involved imperceptible "hereditary factors." However, his approach to nature was so uncommon at that time that his findings went unnoticed for decades. It was not until the turn into the 20th century that he was discovered and his way of approaching nature became the basis for a new scientific discipline that would soon carry the name "genetics."

This new discipline was reductionist in nature. Which means that its focus grew narrower and narrower with time: moving from the whole organism to specific traits, to the microscopic world of cells, to the cell's nucleus, and to its subunits, the chromosomes. From there it postulated invisible components of the chromosomes, which were given the name genes. In this block we follow how this approach to understanding heredity took its course over time. Among other things, the students learn about T.H. Morgan's "fly room" at Columbia University, and how tens of thousands of experiments on fruit flies (*Drosophila*) eventually led to chromosome maps that appeared to be ironclad in their depiction of where specific genes must be located on fruit fly chromosomes.

But, alas, as the history of science so often shows us, there was more to the issue than those researchers realized at the time. With the discovery of pleiotropy and polygenes, the picture of genetic processes became—once again—more fluid and dynamic. Soon thereafter biochemists joined the search for the secret to heredity. Here too, what appeared at first to be an obvious conclusion—that proteins (enzymes in particular) must be the key genetic substance in chromosomes—gets called into question by a small group of researchers who considered a different substance, DNA, to be the key player. All this brings us to the dramatic events that took place at Cavendish Laboratories in Cambridge and at Kings College in London. These eventually result in the discovery of the Double Helix, the description of "the Genetic Code," and culminate in what Francis Crick called "the Central Dogma." On that foundation, the Human Genome Project came about, which was thought by many to be "the solution" to many heredity-based issues.

But alas, once again, the tables are eventually turned. We follow how in the 21st century a very different picture arises, one that is much more dynamic and holistic in nature than the views that dominated genetics a few years prior. Steve Talbott (2010) summarizes the situation as follows:

> *The fixity, precision, and unambiguous logical relations of the code seemed to guarantee its strictly mechanistic performance in the cell. Yet it is this fixity... that has lately been dissolving more and more into the fluid, dynamic exchange of living processes. Organisms, it appears, must be understood and explained at least in part from above downward, from context to subcontext, from the general laws or character of their being to the never fully independent details. In the end, we can meaningfully apprehend the lowest-level activities only so far as we recognize them to be performances of the whole organism.*

With these findings, the questions that we have been exploring in this block, from embryology to genetics, come full circle. The students are able to see how fundamental issues that have occupied biologists for centuries—in particular ones dealing with the relationship of the parts to the whole, the whole to the parts—are still very alive and still essential when it comes to understanding living organisms.

Part III. Ecology

Another dimension of the dynamic interweaving between whole and parts exists in the interaction of organisms with each other and the environment in which they are embedded. This topic is typically covered in a second 11th grade life science block that focusses on ecology.

In our time, all such matters involving the environment are embedded in the increasingly urgent questions around climate change and the role that humans play in it. For 11th graders this issue has been raised in many of their previous blocks. In their 10th grade geophysiology/meteorology block, for example, they have engaged many central questions around climate change. In that context they will have learned about the increases in global surface temperatures in the past half century and how greenhouse gas concentrations have risen constantly since the advent of industrialization and continue to do so. In conjunction with their study of ocean currents—and their effects on global heat distribution—they will have learned about ocean warming, ocean acidification through CO_2 absorption, sea ice melting, coral reef bleaching, and rising sea levels. They will have noted how in recent years extreme weather-related phenomena have become more and more common: exceptional heat waves, drought, wild fires,[3] flooding, etc.

In connection with their studies of the carbon cycle the students have become very familiar with the most prevalent views of how to address the challenges of anthropogenically-induced climate change that involve reducing greenhouse gas emissions by phasing out fossil fuels and moving to various forms of renewable energy. In the 11th grade ecology block we address—among many topics—another level of decarbonization that has a huge significance for biodiversity. The students learn not only that we need to leave more carbon (coal, oil, and natural gas) in the ground, but that we need to leave it in the biosphere as well—and to increase its levels there. What that means is explored concretely toward the end of Part III.

3 For example, *The New York Times* (Sept. 22, 2022) reports that researchers have found a 27-fold increase over the past decade in the number of people experiencing an "extreme smoke day"—which is defined as air quality deemed unhealthy for all age group—caused by wildfires.

in conjunction with the topic of endangered rainforests and the groundbreaking "Andean, Amazon, Atlantic Corridor Project" underway in South America.

Part III begins with an introductory section that draws on the writings of ecologists Aldo Leopold, Holmes Rolston, Karen Warren, and others, and introduces Rolston's idea of "value-ability" as the key to an ecologically sound relationship to nature. This is followed by a description of key moments in the gardening and agriculture curriculum of most Waldorf Schools that plant the seeds for "value-ability" in the students and provide the soil for a lively ecology block in the 11th grade.

With such reflections as a background, we move to examples of possible ecological themes for such a course. The first of these is an introduction to the three primary biomes found on the continent of Africa and how they are intimately related to the Intertropical Convergence Zone (ITCZ).[4] With this meteorological dynamic as a context, the students learn in considerable detail about, and compare with each other, the tropical rainforest of the Congo River basin, the Saharan desert, and the African savannas. Relating the verdant life of the rainforests to the mineral-rich deserts—where all plant life is held back—the students discover how the rhythmical movement of wet and dry seasons across the African savannas mediates between those extremes.

After this introduction to biomes and biospheric rhythms, we move into more detail with several examples of ecological interdependencies,[5] which lead us to the concept of "mutualism" and the realization that all living beings find themselves in symbiotic relationships with other organisms and their surroundings.

With this new perspective informing the way we look at ecosystems, we shift our attention again to biome patterns and discover something reminiscent—although on a much more modest scale—of what we found in Africa. We find running North to South across the center of the United States remnants of an enormous presettlement Oak Savanna. To its west we find tall and short grass prairies that stretch all the way to the Rocky Mountains; to the east we come upon dense forests that run all the way to the Atlantic Ocean. We introduce the students in some detail to each of these ecosystems, which helps give them a sense not only for the ecological relationships that informed a large part of North America prior to settlement, but also for those still prominent in many places today.

4 Most classes will have learned about the ITCZ in their 10th grade meteorology block.

5 For example, the role that dung beetles play in transforming 250 million large manure piles in Australia every day.

From there we look at several variations on this theme, such as the way vegetation changes when we move from the equator to the arctic. This is then compared with how the plant world changes moving upward from the base of a mountain all the way to its icy peak.

After considerations of this kind, we introduce the idea of ecological succession and describe the significant role it played in the birth of environmentalism in this country.[6] This is followed by an introduction to the all-presence of microhabitats, which can awaken the students to a host of hitherto unnoticed phenomena.

Part III concludes with an examination of one of the greatest environmental crises of our time: the devastation of huge expanses of the Amazon Rainforest for logging, mining and agricultural purposes. After describing the degree of destruction and its impact on the global climate, we look at the value of these forests in and of themselves as the most diverse terrestrial ecosystems on the planet. But the issues don't stop there. It is very important for the students to be made aware of the difference between merely protecting an environment (conservation) and rejuvenating it—returning it to the kind of vitality and diversity that it once had, and maybe even beyond that.

This question takes us to the amazing discovery—still unknown to most people—that the lush and bountiful Amazon region did not come about by itself. Human activity over thousands of years played a key role in shaping the biodiversity, distribution of wildlife and environmental health in Amazonia. After describing how this human influence was first discovered in the late 20th century, we move to a concrete example: the ecological wisdom and practices of the Kayapo, an indigenous people of the southwestern Amazon. Through very refined methods of agro-ecology that they have practiced for centuries, the Kayapo have been able to alter and manage ecosystems in ways that have significantly increased their biodiversity.

After going looking into some detail about the Kayapos' ways of working, we move to the story of anthropologist Martin von Hildebrand, who has devoted his life to fighting for the rights of indigenous peoples and has helped them take back control of millions of acres of Amazon Rainforest. In his "Andean, Amazon, Atlantic Corridor Project" he has partnered with indigenous communities, NGOs and national governments to create world's largest ecological corridor, that stretches from Columbia across Venezuela and Brazil to the Atlantic Ocean. This project

6 See Appendix III for a concrete example of ecological succession.

gives the students a sense for whole new possibilities that involve not only stopping ecological devastation, but of developing—with the help of indigenous wisdom and commitment—new ways of rejuvenating the environment and influencing climate change. These concluding topics speak strongly to students who are seeking in their youthful idealism for large and worthwhile goals that can help heal and renew our challenged planet.

Appendices

In addition to the main chapters of this book, I have included three appendices that supplement and expand on certain aspects of the topics covered above.

Appendix I.

"Human Embryology" adds an additional dimension to the opening chapter on developmental biology. Originally written for a collection of essays on human sexuality,[7] it provides a description of human development—from gametogenesis to fetus—and provides numerous pedagogical perspectives on how such a course can help students develop capacities for understanding human development—and living organisms in general—in a much more dynamic and feeling-permeated way.

Appendix II.

"Colorado High Country—A Comparison" brings an example of how a teacher can select out certain aspects of an ecosystem and go into great detail using what Rudolf Steiner sometimes called "the comparative method." It explains how this approach allows phenomena to characterize each other and goes on to illustrate this by comparing two prominent tree species that populate many high-altitude valleys and ridges in the Rocky Mountains of Colorado.

Appendix III.

"Investigating a Local Ecosystem with Students - an Example" takes the comparative method to a new level by investigating the changes in vegetation when one moves from a sandy beach at the edge of Lake Michigan into a well-developed Oak savanna further inland. Using non-phylogenetic plant classification methods and applying them to changes in Plant Available Light (PAL), Plant Available Moisture (PAM), and Plant Available Nutrients (PAN), we witness a very tangible example of ecological succession—similar to the discoveries of Henry Cowles 125 years prior—as well as the pedagogical value of such investigations.

---

7 *Trailing Clouds of Glory* (2014). Douglas Gerwin (ed.). Waldorf Publications, pp. 250–270.

PART I.

Developmental Biology

Getting Started

It's a very odd thing,
As odd can be,
That whatever Miss T eats,
Turns into Miss T.[8]

We heard about Miss T when discussing the immune system in the 10th grade, but the mystery that surrounds her has not yet been fully vetted. That mystery can be addressed in numerous ways. One possibility is to gather with the students a few characteristics of organisms as we know them. Everyone recognizes, for example, that animals and humans grow, develop and age, that they eat, breath, defecate, reproduce, and eventually die. While this is common knowledge, it sometimes helps to follow such observations with a comparison. For example: How does a dog differ from a lawn mower? It is rapidly clear that what was just enumerated applies to dogs, but not lawnmowers. What's more, we can construct lawn mowers, but we can't construct dogs, or frogs, or a horse.

One of the central questions for developmental biology is how organisms are able to take inanimate material and build it into a living, dynamically-developing bodily organization. This field of study addresses many riddles that non-biologists tend to take for granted in everyday life, but shouldn't. For example, starting as a fertilized human egg cell one-tenth of a mm in diameter, we find at the end of nine months a fully developed and functioning organism that now consists of a community of some 30 trillion (30 million million) cells.[9] We can let the mathletes in the class figure out how rapidly cell production has to take place for that to happen. They find that there have to be—on average—a production of around 110 billion cells/day (which would mean ~4.5 billion cells/hour, ~75 million cells /minute, ~1.25 million cells/second), to produce the full set of 30 trillion in nine months.

Even more amazing is to realize that those are not 30 trillion repeats of the original cell. Au contraire, they have morphed into a very differentiated community of some 200 cell types that make up a wide range of tissues and structures: skin, muscle, liver, kidney, etc.—all working together in a highly coordinated fashion in service of the organism as a whole. Out of this mindboggling phenomenon, several

8 Walter de la Mare—"Peacock Pie."
9 Some estimates go as high as 100 trillion cells at birth.

additional riddles that we will be pursuing in this block can be highlighted for the students:

- During the nine months before birth, the cells do not only divide, but—as just indicated—they differentiate. One central question for developmental biologists is: How does a single cell give rise to hundreds of different cell types.

- These cells do not only differentiate. As they develop they become organized into tissues and organs, in the course of which they migrate, they die, they fold and they separate. Such processes—known as morphogenesis (the genesis of form)—result in a highly ordered composition in which, for example, the eyes end up in the front of the head (not the back) and on both sides of the nose (not above or below it), the fingers—not the toes—are located at the ends of the arms, and so on.

- Even the seemingly straightforward process of growth through cell division presents a significant riddle. For example: Why do cells stop dividing when they do? If each cell in someone's face divided just one more time, they would be considered dreadfully malformed. If someone's arm cells went through just one more round of cell division, they would be able to tie their shoes without bending over, and basketball would become "a slam dunk"! Again: What determines that the cells stop dividing when they do?

Organisms Unfolding

Epigenesis or Preformation?[10]

How does the fertilized egg (zygote) develop into an organized whole with multiple harmoniously collaborating organs? One way to get the students thinking about this question is to describe how, at the dawn of Greek medicine over 2000 years ago, two very different explanations arose for how the human organism could arise out an egg cell.

Aristotle thought that the differentiation of the organs emerges gradually and that they are formed out of the undifferentiated material of the egg cell. This view was given the name "epigenesis." Epicurus, on the other hand, believed that the egg contained a completely formed child in miniature, one who simply grew larger with time. A famous Roman, Seneca, formulated this "preformationist" view as follows (cited in Kennedy, 1967):

> *In the seed are enclosed all the parts of the body of the man that shall be formed. The infant in his mother's womb hath the roots of the beard and hair that he shall wear some day. In the little mass, likewise, are all the lineaments of the body and all that which posterity shall discover in him.*

The debate between these two perspectives lived on for two millennia, although by the 18th century the preformationist view appeared to be the victor. It existed in two different versions. The "ovists" maintained that the preformed human was in the egg of the mother. This had been the assumption up until the advent of the microscope, when one preformationist thought he saw a tiny human being with head, body, feet and hands in each sperm (animalcule)! An illustration of a microscopic infant folded in a fetal position within the head of a sperm—made in the 17th century by Nicolaas Hartsoeker—can help the students better picture what is being imagined here.

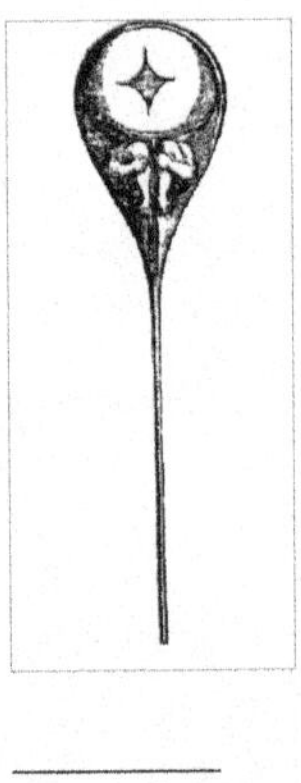

Fig. 1.1 Preformed infant in head of sperm. (Nicolaas Hartsoeker, Wikimedia Commons)

10 References to the resources used in each short section of the text—which usually begin with their own heading—can be found in parentheses at the end of the section. Direct quotes are cited in the text itself.

The sperm-based "animalculist" perspective created a great split between the animalculists and the ovists, because it suggested that Adam rather than Eve contained all of humankind within himself.

Although most 11th graders do not consider either of these preformationist explanations to be self-evident, the split between them brings forth one necessary—and very hard to imagine—consequence for the interpretation of human development: If the embryonic egg—to take the ovist perspective—contains a complete miniature body with all its organs and parts, then it must also contain (if female) the proportionally tinier eggs of the next generation, which themselves must contain entire bodies containing even smaller eggs (if female) of the next generation, an so on and on (and smaller and smaller). In this way, all of future humanity had to be existent in Eve—preformed but not yet grown large—in a manner something like a never ending and ever smaller series of Russian dolls.[11] [12]

So how can the dichotomy between preformation and epigenesis get resolved, we ask the students? How do we get beyond these two very different forms of speculative thinking and determine what is really going on? For a person living in the 21st century the answer is, of course, easy. We have to observe the egg (ovum) at

Fig. 1.2 Russian dolls. (Kora27. CC BY-SA 4.0 International)

11 Although the size gaps between the generations would be much greater than the gaps that exist between Russian dolls.

12 If we stop for a moment and compare the two ways of thinking that dominated the understanding of development up until the 19th century, we can see how one—epigenesis—approaches the world as process, as unfolding in time; the dynamic and flow of existence catches the eye; it is attuned to the "verb" side of nature. The other perspective—preformation—focuses on finished forms; it emphasizes discrete features as they manifest in space. The world appears to this way of looking more like a noun than a verb. Although it is far too early in the block to emphasize how these two "ways of looking" are fundamental in the unfolding of modern biology, as teachers it is fruitful to keep this polarity in mind as we move on, paying attention to how it reappears again and again in new forms as we progress from the 19th to the 20th and then into the 21st century.

the beginning of its development, and pay attention to what happens from there on out. This was, however, easier said than done in the 18th and early 19th centuries. On the one hand, there was the need to break free from the reliance on past authorities, both religious and scientific. On the other hand, this particular question required the help of developing technology. It was not until the 1820s, when a combination of improved microscopes, new staining techniques and institutional changes in German Universities, led to an overcoming of the preformationist view and the birth of a new field of study: descriptive embryology.

Although compound microscopes had already been invented in the late 16th century, the early microscopes suffered from chromatic aberration, causing colorless objects to shimmer with the colors of the rainbow and making exact observations difficult. It was not until the early 19th century that achromatic lenses were developed, making accurate micro-observations of organic tissues possible for the first time. Already the first year that the new lenses went on the market, in 1827, the Estonian Ernst von Baer was able to make a huge discovery. Up until this point, the mammalian egg (ovum) had never been observed. The large eggs of birds, reptiles, fishes, and amphibians were well known, but in mammals they were still veiled in mystery. Some even wondered whether they actually existed. One day, however, while investigating dog ovaries with the naked eye, von Baer noticed "a whitish-yellow point which was in no way attached" and "swam freely in its liquid." He removed it, placed it under one of the new microscopes and could hardly believe what he had found. "I recoiled as if struck by lightning…. I had to try to relax a while before I could work up enough courage to look again, as I was afraid I had been deluded by a phantom. Is it not strange that a sight which is expected, and indeed hoped for, should be frightening when it eventually materializes?" (Gilbert 2010, p. 16; Moore 2003, p. 404)

This was a truly historic moment. Since the early Greek culture, female humans had been thought to carry eggs from which their children develop (preformed or not), but no one had ever knowingly seen such an egg.[13] And now, for the first time, von Baer confirmed through technology-assisted observation the expected—but at the same time frightening—"fact" that such mammalian eggs do, indeed, exist. A profound turning point in the development of the life sciences had been reached. With the help of technological innovations, a subsense-perceptible level of existence had become accessible for scientific study, thereby shifting the focus of

13 Although earlier claims were made in 1797 and 1824, von Baer was the first to describe and clarify some important details and is thus given credit for being first to have actually observed a mammalian egg.

much research, and bringing with it an increased emphasis on reductionistic causal explanations.

So what happens now? We would expect the study of mammalian embryology to flourish. After all, people have been asking the question: preformation or epigenesis? for over 2000 years, and now, finally, the tools were available by which the questions could be answered.

Interestingly, however, when von Baer read his paper describing this discovery at the Annual Congress of the Society of Natural Scientists and Physicians in 1828, it was received in silence. The researchers there could not yet imagine how this discovery would provide the empirical basis for the next big step in an emerging field of scientific investigation.

As is often the case in the real world of scientific research, the next step was more difficult than we might think. There were, for example, concrete factors that prevented the research from moving "full steam ahead." For one, mammalian eggs are extremely tiny (the human zygote is only about one-tenth of a mm in diameter). Although it is precisely here where the microscope is able to help out, there were additional challenges. For one, mammals only produce a small number of eggs, which limits the amount of material available for study. But an even bigger hurdle—one which the students should be able to come up with themselves—is that mammalian embryos are not available for study in the external environment. They only exist inside the mother organism, where they cannot be observed directly! Summing all that up, it became probable that most of the pioneering research on embryological development would take place on non-mammals. Indeed, most biologists took the easier option of studying embryonic development in creatures whose eggs were relatively large, plentiful in number, and developed outside the mother organism. Amphibians, in particular, became a favorite object of study.

Before diving into the amazing findings revealed by the study of amphibian embryos, however, it is helpful to provide the students with an overview of basic developmental phases using the Lancelet (*Branchiostoma floridae*) as an example.[14] The Lancelet, similar to mammals, has an egg in which the yolk is rather evenly distributed (isolecithal),[15] which provides an easily overviewed instance of key

14 The Lancelet belongs to the chordate subphylum Cephalochordata. It is a small, translucent, fish-like animal found in shallow marine waters. In contrast to the vertebral column of the fish, its main supportive structure is a highly developed notochord that lies below the nerve chord and dorsal to the intestine.

15 Although, because yolk is heavier than ordinary protoplasm, it tends to be more plentiful in the lower part of the egg.

developmental phases common to all vertebrates. (Francis 2011; Gilbert 2010; Gould 1977; Gray 1967; Hays 1972; Hickman 2001; Lewontin 2001; Marienschein 2017; Moore 2003)

Early Development of the Lancelet (Branchiostoma floridae)

To start off, it is interesting for the students to contemplate how the round zygote will go about changing from its spherical form into the more complex, linear shape, we find in the adult Lancelet.[16] Does something head-like start protruding out in the front, and something tail-like begin to stick out in the back, or what?

Rhythmical Cell Division (Cleavage). After some discussion of how bodily forms might arise, the teacher can begin drawing on the board and describing how we have to wait about 8 hours after fertilization before the first visible changes begin.[17]

A short groove appears at the upper pole and gradually lengthens to create the first cleavage furrow, which forms two cells (blastomeres) from the original one. A second cell division starts at about 12 hours, resulting in a 4 celled organism. Once again, its plane is vertical and at right angles (seen from above) to the first cleavage. The third division (8 cells) begins at about 18 hours and is horizontal, with its cleavage plane above the equator. These divisions continue rhythmically: 16 cells at ~21 hours, 32 cells at ~24 hours, 64 cells at ~27 hours, and so on. Only after about 36 hours, at the onset of the early blastula phase, does the cell division become asynchronous. (The students also find it fascinating to learn that if a group of eggs are fertilized at the same time and in the same environment, there is an amazing synchronization in their development. Each new cleavage begins at precisely the same time in all embryos. In fact, through all of their early development they reach stage after stage at almost exactly the same time.)

It is interesting to note that with each cell division the cells become smaller and smaller, but there is no increase in the overall size of the embryo. Why not, we ask the students? Because up to this point no new nutrients have been taken in. What we are observing is a process of differentiation within the original whole, but with no "add-ons" or actual growth. In fact, the dry weight of the embryo decreases slightly due to the metabolic activity required for the cleavage process.

16 Students who already have learned this are asked to "keep the secret" for the moment.

17 The developmental timing that will be given here is approximate, since the pace of such events varies greatly with changes in water temperature.

After gathering the first observations described above, it is fruitful to pause and go back through what has been described, letting the students point out some of the details and in the process confirm which characteristics are essential and which are perhaps a product of the teacher's limited drawing ability. For example, the first two cleavages are vertical and extend from pole to pole. The third (8 cell stage) is horizontal. Is it always like that? Yes, we confirm. Then comes a vertical cleavage (16 cells), followed by a horizontal one (32 cells). We explain how this process continues (64,128...), but gradually becomes less synchronized as the number of cells grows larger. Around the 16- or 32-cell stage, the embryo begins to look a lot like a berry and is given the name "morula" (Latin: morus = mulberry).

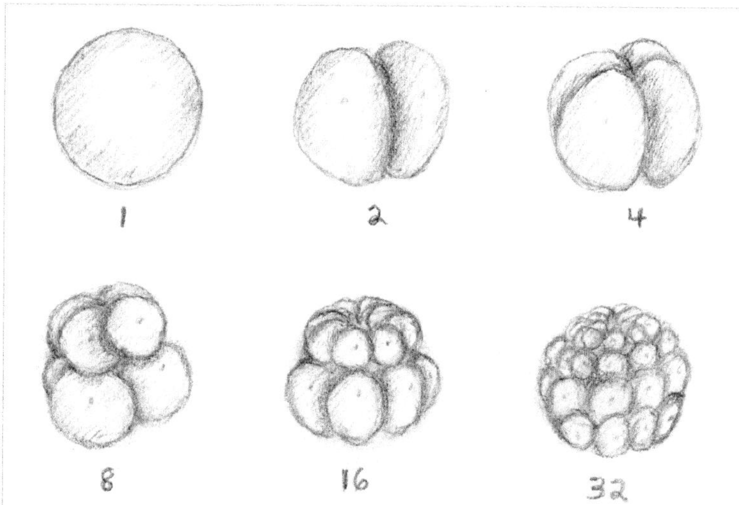

Fig. 1.3 Rhythmical cell division (cleavage) of lancelet egg cell during the first 24 hours after fertilization.

The Blastula.[18] As the number of cells increase, contact with the nonadjacent cells is no longer possible and a fluid filled cavity develops called the blastocoel. The embryo is now called the "blastula." We look again at what is on the board. Do we see anything else noteworthy, or questionable? What about the first horizontal cleavage. It's not quite in the center. Drawing malfunction or actual phenomenon? It's the real deal, and we notice that this has consequences: It results in larger cells in the lower hemisphere and smaller ones in the upper. Which is to say, the larger cells below contain more yolk that those above. The upper hemisphere soon acquires the name "animal pole," and the lower "vegetal pole." These names are not arbitrary, but we have to learn more before we can understand why.

18 Onset at about 1.5 days.

Gastrulation.[19] The board drawings continue and we see how a new major event begins when the vegetal pole starts to flow into the blastocoel (invagination), resulting in a double-layered hollow sphere with the smaller cells of the original animal pole on the outside and the larger, yolkier cells from the vegetal pole lining the inner surface. (The whole process reminds one of the "L" gesture in eurythmy.) This process is known as gastrulation. Its end product is called the gastrula, and the hollow space the gastrocoel.

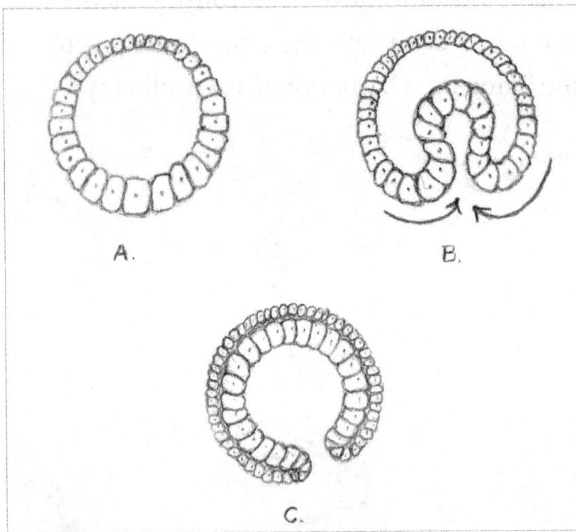

Fig. 1.4 Frontal section of the gastrulation process.
A. Blastula. **B.** Invagination of the vegetal pole.
C. Gastrula with two layers: ectoderm and endoderm.

Neurulation.[20] Further drawings show how the outer layer of the animal pole begins to infold (invaginate), forming the neural tube, while below it, the inner layer of the gastrula begins to fold outward (evaginate), forming lateral pouches that later meld to form what will become the body cavity (coelem). Between the laterally outpouching tissue, yet another upward folding (evagination) takes place, forming the notochord. Emphasizing the creation of the neural tube, the outcome of this phase of development is called the Neurula.

A. The ectoderm of the animal pole begins to infold (invaginate). B. Further invagination of the animal pole. The inner endoderm layer begins to fold outward (exvaginate). C. With neural tube and somites almost fully formed, additional out

19 Onset at about 3.5 days.

20 Onset at about one week.

and upward folding of endoderm, which will form the notochord. D. Neurulation completed with the new middle layer (notochord and somites) consisting of mesoderm.

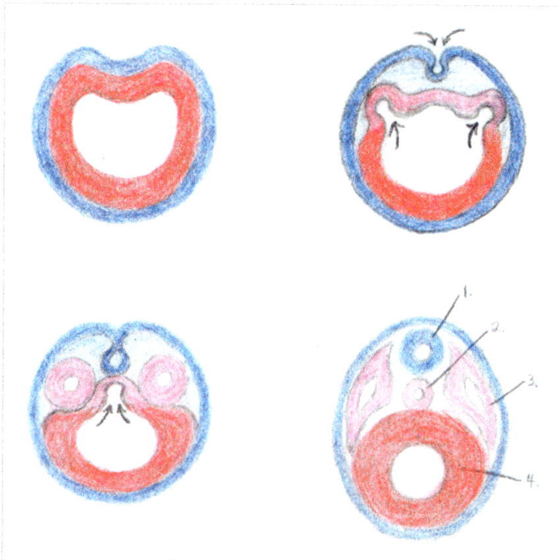

Fig. 1.5 The neurulation process. Blue: Ectoderm. Red: Endoderm. Pink: Mesoderm. 1. Neural tube. 2. Notochord. 3. Somites. 4. Primitive intestine.

The Three Germ Layers. As a result of these outpouchings and infoldings, we are now able to identify the three layers, which appear early in the development of all vertebrates. We color code them in the drawings and explain that the "ectoderm" is located on the outside (+ neuraltube), "endoderm" on the inside, and "mesoderm" between them.

The students then learn that after the formation of these three germ layers, the cells in each layer begin to differentiate in specific directions. It shouldn't surprise them that the outer layers of the developing Lancelet, the ectoderm, will differentiate into skin (epidermis), hair, the nervous system and sense organs. The endoderm, by contrast, will become the organs of the digestive tract and breathing. The mesoderm will develop into the skeleton, muscles and the circulatory system. In other words, all the organs that will eventually develop can be traced to these three layers. (The three germ layers were first discovered by von Baer and his colleague Pander in chicken embryos. Von Baer was so amazed by this finding that he looked for something similar in as many other creatures as he could—mammals, fish, reptiles—and not only found the same three germ layers, but that in each instance the same organs developed out of the corresponding germ layer: the heart formed out of the mesoderm, the brain out of the ectoderm, and so on).

It is fruitful to take a step back at this point and look with the students at the overall gestures of what we have seen unfolding. Several things are notable and can take the discussion in several directions, but with the help of the teacher, it is valuable to include the following aspects.

It is striking to observe that we start out with a round, seemingly amorphous whole, to which nothing is added. If we think back to our earlier discussion of building a bicycle, it is clear that embryonic development is not put together piece by piece. Instead, it begins to divide itself into smaller and smaller units. In the process an above and below becomes evident: The larger, yolkier cells settle at the bottom, the smaller, leaner ones above. The whole process is very rhythmical and it seems at first as if it could go on the same way forever (2 —> 4 —> 8 —> 16, and so on). It is important to emphasize, too, that the embryo has not increased in size, but has only been subdivided into smaller and smaller cells.

But then, suddenly, what had been taking place on the surface starts to flow inward from the vegetal pole. After this infolding we see further development from the inside out, as well as a new impulse of infolding from the top (animal pole) downward. What does this tell us about characteristics that this creature will later possess? Will it take in the outer world as a fully developed organism? When we relate that question to what we see happening here, the students recognize that an animal takes in food and air in order to live. Looking at our embryo we see that when the vegetal (vitality) pole folds in it is starting to form what will become the digestive and respiratory tracts, which are the life-sustaining organs through which the organism brings the outer world (food and air) into itself.

On the other hand, the smaller, leaner ectoderm cells of the animal pole (Latin: animus = mind, soul) fold in and create organs (senses, nerves, brain) by which the organism "takes in" the world consciously. They are the basis for the inner life, the means by which the outer can be experienced inwardly.

And, of course, uniting these two extremes we find the circulatory system that develops out of the mesoderm. Through the skeleton and muscles, the mesoderm also provides the structure and the potential for movement within the whole. After the formation of the three germ layers, the students are often surprised to learn that a new and quite different kind of the process begins: cell differentiation. The cells begin to differentiate in conjunction with the specialized roles that they will play within the tissues and organs where they are located—the cells in the walls of the intestine, for example, will need to be very different from those in the gills or in the brain. (We will take an extensive look at the interesting topic of cell differentiation later in the block.)

Before moving on, we provide a side-view sketch of the Lancelet. This complements the one-sided frontal section perspective we have taken so far, and gives a sense for the entire organism that will result from further differentiation.

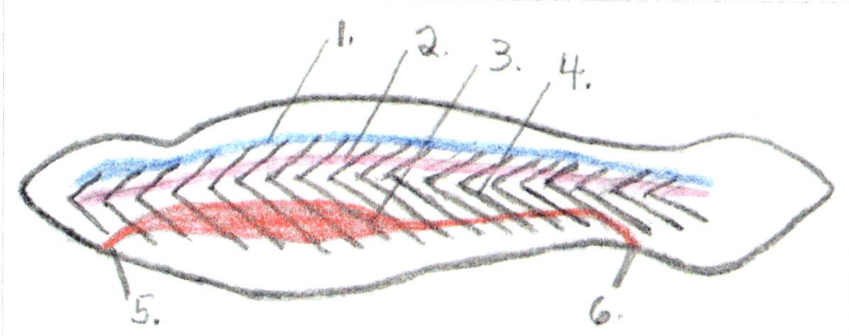

Fig. 1.6 Simplified lateral section of lancelet. 1. Neural tube. 2. Notochord. 3. Digestive tract. 4. Segmented muscles. 5. Mouth. 6. Anus.

Even though the Lancelet is a "mere" Cephalochordate, what we have just covered provides the students with an introduction and uncomplicated overview of the first phases of vertebrate embryology, as well as some of the basic concepts used in describing embryonic development. In the process it became clear that epigenesis, not preformation, best describes what is happening on the path from egg to adult organism. Observations such as these are representative of findings from the field of descriptive embryology, which was the primary form of embryology in the half century following the development of the achromatic microscope. For the first time in history, discoveries of this sort were taking place that pertained to a whole range of organisms. This exciting phase in the history of biology kept embryologists very busy for a long time. (Gilbert 2010; Hurney 2015; Moore 2003; Romer 1997; Shubin 2009)

A New Approach: Experimental Embryology

It was only in the last quarter of the 19th century that a change of emphasis began to appear. The new approach was given the name "experimental embryology" to indicate that the experimental mindset that had grown popular in so many realms of science was also now finding a home in embryology. Researchers were now looking for more than descriptions, they were seeking causal explanations.

They did this by disturbing in various ways the normal course of embryonic development. Through invasive experiments, they sought to discover the laws of causality at work as organisms unfold. We remind the students how this kind of exploration starts with a hypothesis, which is then translated into an experiment, the results of which will either support or run counter to the scientist's expectations and predictions.

In its early days, two schools of thought strongly influenced research in this new field of study. The one group proposed a "mosaic theory," where the egg is seen as divided up into many tiny regions (like a mosaic), each of which is destined to produce specific parts and organs of the adult organism. Development (ontogeny) is merely the unfolding of a predetermined structure. In the words of one of the originators of the mosaic theory, August Weismann, we should picture a fertilized egg in which the starting points (primordial) of organs and tissues "stand side by side, separate from each other like the stones of a mosaic, and develop independently, although in perfect harmony with one another into the finished organism" (Gray 1967, p. 18).

The other school of thought spoke of "regulative development." It assumed that egg development involved a certain flexibility, an unfolding that could manifest in various ways depending on developmental conditions. Each part (area) was seen as regulated by the whole and thus flexible when faced with changing conditions.

After hearing about these two views, there are often students who comment that this dichotomy reminds them of the dispute between preformation and epigenesis discussed earlier. If so, it makes sense to discuss what this dichotomy and the earlier one actually have in common and what they imply about the overall nature of organisms. If the students don't bring it up themselves, such a discussion can easily wait until later.

At this juncture, a stimulating discussion can ensue by asking the students how they would test these two theories. After some discussion, we can share what the experimental embryologists actually did to verify or negate these contrasting viewpoints. One mosaic theory supporter, Wilhelm Roux, formulated the following hypothesis: "If the plane of first cleavage divides the embryo into a right and left half, each half must contain the primordia (embryonic starting points) for that specific half. Therefore, the destruction of one cell at the two-cell stage would also destroy the primordial for half the body." (Moore 2003, p. 449)

In order to test this hypothesis Roux operated on over 100 frog embryos at the two-cell stage by inserting a hot needle into one of the blastomeres and killing it. In about 20 percent of the embryos the undamaged half survived the operation. In some of those cases, the undamaged blastomere developed a half-blastula, neural fold, and mesoderm layer. In other words, the beginnings of a half-tadpole developed on the undamaged side. Roux concluded therefore, that each side of the egg was predetermined to become that part of the organism, thereby appearing to confirm (or at least support) the mosaic theory.

For scientists working in the rapidly developing field of experimental embryology, these findings were a "big deal." In the spirit of the modern scientific method—which calls for experiments to be replicated by independent researchers in order to confirm the reported findings—other researchers were eager to repeat Roux's experiments. There was a rush to the laboratories where researchers who had previously been focused on observing how life unfolds, were now filled with the impulse to experiment and test the theories that were being hotly debated. It was noted by some, for example, that in his experiments Roux allowed the deceased cell to remain attached to the living one. They raised the question of whether this might have influenced the development of the undamaged side. In pursuit of an answer to that question, one researcher in particular, Hans Driesch, made some very interesting discoveries using two-celled sea urchin embryos. We will, however, save Driesch's findings for later, and move first to Hans Spemann and the historic Nobel Prize winning discoveries that he and his team of graduate students made while experimenting with amphibian embryos. (Fels et al 1976; Francis 2011; Gilbert 2010; Gould 1977; Gray 1967; Moore 2003)

Spemann's Newt Experiments

Prior to the turn into the 20th century, a handful of leading figures in the new field of experimental embryology had performed experiments on a variety of animals. Their results were, however, very few in number, difficult to compare, and in some instances even contradicted each other. Their methods were also dissimilar and quite unrefined.

In the early 20th century, Hans Spemann became the leading figure in this new field, not only due to the results he attained, but because he was very clear about two preconditions necessary for accurate and verifiable outcomes. First of all, Spemann realized that it was necessary to focus one's research on one specific creature and to be very thorough in investigating it. In that spirit, once Spemann chose newts (*Triturus cristatus* and *Triturus taeniatus*)[21] as a suitable object of study, he stuck with them. Secondly, Spemann understood that when dealing with such sensitive organisms as developing newt embryos, it is necessary to use very exacting experimental methods that can be reproduced by others. He realized that such experimentation not only requires great manual dexterity, patience, and the willingness to discontinue an experiment when it is not succeeding, but that it also demands equipment that is fine-tuned to the task at hand. At that time, equipment suitable for operating on newt embryos did not exist. Thus Spemann undertook the painstaking process of developing the tools needed. He was able to create a whole toolkit of minute instruments out of glass that were suited to this particular task. (Fels et al. 1976; Gilbert 2010)

In what follows, I will present six classic experiments out of the many performed by Spemann and his students over several decades. (I usually give each experiment a discrete number—as I have done in what follows below—which makes it easier for the students to keep the experiments straight and to facilitate unambiguous references to them during discussions.)

Spemann: Experiment I

Spemann already needed his capacity for innovation at the outset of his research program. In the attempt to test Weismann's and Roux's mosaic egg theory, he decided to "lasso" a newt egg in its first plane of cleavage (the two-cell stage) and pull the lasso tight, which he hoped would separate the two blastomeres completely. But there was a problem—as there often is in real-life experimentation. His team couldn't find a wire or thread fine enough to perform the task without injuring the cells while separating them. But then Spemann had an "aha" moment. The hair of his infant daughter Margaret was soft, thin, and very pliable—just what was needed for their experiments. So, Spemann snipped (plucked?) a few strands from the head of his darling daughter when she wasn't looking and took them to the lab, where he hoped they would perform the task perfectly. But alas, even Margaret's ethereal locks were still not fine enough. But again, an "aha moment" rescued him: He realized that he could de-fat the individual strands and make them even finer. It worked! The defatted lasso was looped along the cleavage line, slowly pulled tight, and "basta," they now had two freshly separated newt blastomeres before them!

21 Newts are semi-aquatic salamanders that are related to frogs.

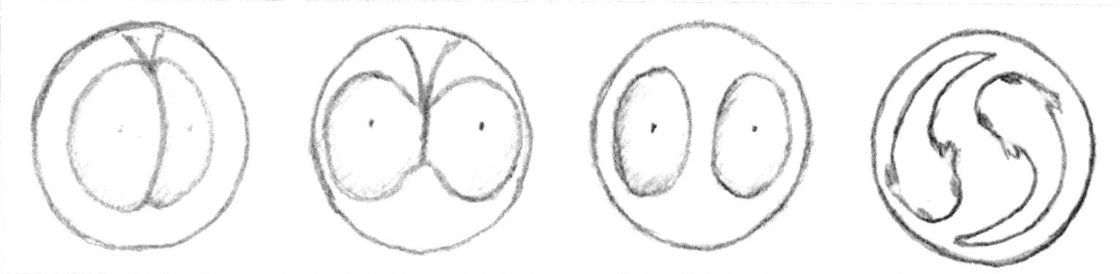

Fig. 1.7 Separation of newt blastomeres at the two-cell stage resulting in identical twins.

We can now ask the students what they expect to develop out of these two cells a) according to the mosaic egg theory, and b) according to the regulative egg theory? The actual results of the experiment—illustrated below—show that the two blastomeres did not develop into two half-newts, as you would expect with a mosaic egg, but into two complete ones—although half size.

The results speak clearly for regulative development. But why are the newts half-size? We explain to the students that all nutrients for the developing embryo have to come from the egg yolk, and until development is complete no other nutrients are available. In that case it is clear, since each blastomere had only half the nutrients available from the original egg to build up its body, it can only be half-size. Based on this result, the students usually want to clarify the difference between identical and fraternal twins. And what about triplets and quadruplets? We tell them how in further experiments, Spemann separated the blastomeres at the four-cell stage and in some instances obtained four quarter-size newts—although this succeeded much less frequently than at the two-cell stage. (Fels et al. 1976; Gilbert 2010; Shubin 2009)[22]

To awaken interest in how Spemann and team proceeded from there on, we ask the students if they have ideas for additional experiments along these same lines that could supplement what was found in Experiment I. Thereafter we proceed with what Spemann and team actually came up with.

Spemann: Experiment II

In another series of experiments, Spemann's lab returned to the newts and began separating the two blastomeres along same the cleavage plane as before, but this time they left a plasma bridge between the two so they would not separate

22 Hans Driesch, who worked with sea urchins in Naples and succeeded several times with this experiment at the eight-cell stage, is said to have succeed once at the 16-cell stage, in which case all 16 blastomeres developed into complete newts (Kennedy 1967).

Fig. 1.8 Partial separation of newt blastomeres at the two-cell stage resulting in conjoined twins.

completely. Once again, we let the students formulate what they think the outcome will be. Then we provide a drawing that shows the result: two newts with fully developed heads and torsos, but with their tail regions still connected. (Fels et al 1976; Nitschmann 1986)

Spemann: Experiment III

Spemann's students and others also undertook the opposite, but much more difficult process. They tried to unite two newt zygotes just as they were beginning their first cleavage by placing the two zygotes crosswise (X) over each other along the cleavage plane. After letting the students propose what they think will happen, they learn that although many did not survive, in some instances the two zygotes merged and amazingly, a giant embryo developed therefrom![23] Why a giant one, we ask? By now the students should know: By merging the two eggs, twice the normal raw material was available for developing the new creature.[24] (Fels et al. 1976; Nitschmann 1986; Schnetter 1968)

Fig. 1.9 Union of two newt zygotes along the first cleavage plane resulting in a "giant" newt twice the normal size.

23 Driesch also performed a corollary experiment by fusing two sea urchin embryos at the blastula stage to produce a single giant larva.

24 Living in Chicago, it can sometimes occur that the students begin (with a little nudge from the teacher) to speculate about whether such an event may have stood behind the creation of one of Chicago's most beloved football players: "the Frig" (short for Refrigerator), who, at 6'2" and 345 pounds, led the Chicago Bears to a Super Bowl victory in 1985.

Prospective-Outcome, Prospective-Potential.

These first three experiments reveal the regulative capacity newt embryos possess—which was soon confirmed by various experimenters working with amphibians and sea urchins. As in all scientific developments, the empirical findings are only half the issue. Just as important are the thoughts developed to place them into a larger conceptual framework. To shed light of the experiments just described, two concepts previously developed by Hans Driesch are helpful. Driesch distinguished between the "prospective-outcome" (PrO) of an embryo (or part thereof) and its "prospective-potential" (PrP). At the two-cell stage, for example, the prospective-outcome of the left blastomere is to become the left half of the newt, and similarly, the prospective-outcome of the right blastomere is to become the right half of the newt. This concept points to what will normally happen, to the fate of various tissues under unaltered conditions. The prospective-potential of those same blastomeres is, however, much greater than what normal development shows us. Both blastomeres in the Spemann experiments had—as experiments I, II, and III show—the potential to become an entire newt under altered conditions.

Once these two concepts have been discussed with the students, we can ask how they relate to the mosaic and regulation theories? It is easy to see, of course, that in the mosaic theory the prospective-potential is equal to the prospective-outcome (PrP = PrO).[25] The regulative egg theory, by contrast, says that the prospective-potential is greater than the prospective-outcome (PrP > PrO).[26] Reflecting further, it becomes clear to the class that the incredible potential found in each blastomere at the two-cell stage cannot remain forever. As they will know, various parts of an adult newt are no longer capable of developing into an entire newt, regardless of the conditions.

With that insight in mind, the class can reflect on what would be the next logical question for a researcher in this field to pursue? Historically there was, in fact, a new and fundamental question awakened by experiments I–III and the concepts of PrO and PrP—one that would lead to a Nobel Prize for Spemann. It went something like this: If early on the prospective-potential in various parts of a developing organism is very large, whereas by the time it reaches adulthood that potential is largely gone, then how does this shrinking of prospective-potential take place? Is it the same everywhere throughout the organism? Does it happen all at

25 According to this view, there is no inherent flexibility in the egg cell, nor in what develops from it over time, that allows for adaptation to new circumstances.

26 Spemann's experiments clearly demonstrate this.

once, or gradually, or how? Once researchers saw answering such questions as the research target, the next challenge was to create with experiments that could answer them. We now give the students the opportunity to come up with imaginative ideas for experiments that could these questions. It is often fruitful to break them up into small groups for this activity. Each group can then share their ideas before we move on to how Spemann and team actually proceeded. (Fels 1976; Gilbert 2010; Linder 1998)

Spemann: Experiment IV

To address these questions, Spemann developed a transplantation technique—and new tools to go with it—that would enable his team to exchange tiny segments of embryonic tissue between two embryos. To picture how challenging this was, the students need to understand that a newt egg is only 2mm in diameter. It is extremely challenging to remove miniscule segments from specific locations in two such tiny embryos and then transplant each piece into the gap created by the removal of the corresponding segment in the other embryo. To this end, Spemann developed very refined transplantation pipettes out of glass that were able, through suction, to draw a small tissue-section into the pipette. The section could then be transported in the pipette and inserted into the opening created by the section removed from the other embryo—and vice versa.

As the students can well imagine, this painstaking task took considerable practice and dexterity. After a time, Spemann's lab members became quite adept at the procedure and were able to move forward with their planned experiments. In one set of experiments, they removed segments of "presumptive stomach skin" (segments whose prospective outcome was to become stomach skin) and exchanged them with "presumptive nerve tissue" (neural ectoderm) located at the dorsal pole of the embryo. This was done at two different phases of development. One group the newts was done during the blastula phase and also at the outset of the gastrulation process. What happened? The presumptive stomach tissue transplanted into a location that would normally become nerve tissue became nerve tissue. Similarly, the presumptive nerve tissue transported into the stomach region became stomach skin. The same result was found at the outset of gastrulation (Group 2).

This kind of outcome was characterized as "location-correspondent development." Clearly the way the transplanted tissue segments developed was determined by their new location, not by where they had come from. In terms of prospective-potential (PrP) and prospective-outcome (PrO), we can say: At

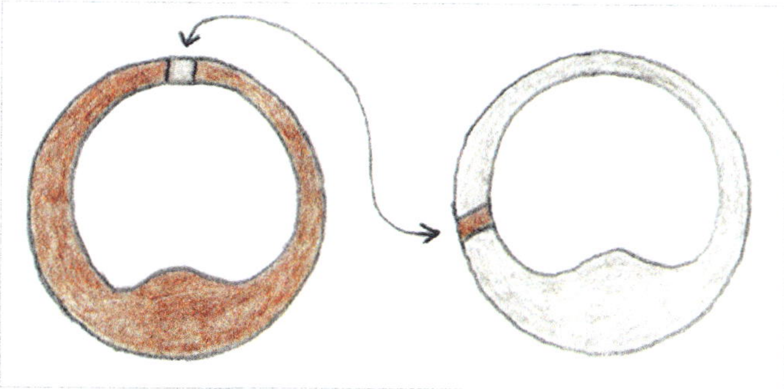

Fig. 1.10 Transplantation during the blastula phase. Presumptive nerve tissue is exchanged with presumptive stomach skin. Both develop in accordance with their new locations.

the blastula and early gastrulation phases of development, the newt tissue is not yet fixed and inflexible in regard to the direction it will develop. Using our new vocabulary, we can say: during these early phases of development the PrP > PrO. At this early stage, newt tissue can adjust to circumstances and develop in accordance with the larger whole of which it is a part.

In my experience, it helps a class if the teacher provides an analogy or two to illustrate what we are talking about here. For example, the teacher can refer to one of the students in the second row, let's call him Horace, and point out that if he had been born in Brooklyn, then his PrO would have been to speak with a heavy Brooklyn accent. But if, lo and behold, his family had moved to Dallas when he was three, he would have soon been able to drawl with the best of them. In fact, he would have quickly lost his Brooklyn accent and would have never gotten it back again. But now, as a teenager living in Chicago, when he asks—with a drawl—why don't "all y'all" wear cowboy boots to school, you can tell that either he is still new to Chicago, or if not, that he has lost most of his "accent & wardrobe PrP," and is unlikely to ever become a state-of-the-art Chicagoan. (If, however, you are teaching this class in Texas, you will probably want to reverse this sequence and have Horace demanding to wear "flip flops" during the summertime, even when he is working on the ranch or climbing mountains.) (Fels et al. 1976; Gilbert 2010; Linder 1998; Nitschmann 1986)

Spemann: Experiment V

Once the results from Experiment IV are in, it is clear that team Spemann is part way there in their attempt to answer the research question formulated above. The next, very logical step, would be to explore the relationship between PrP and

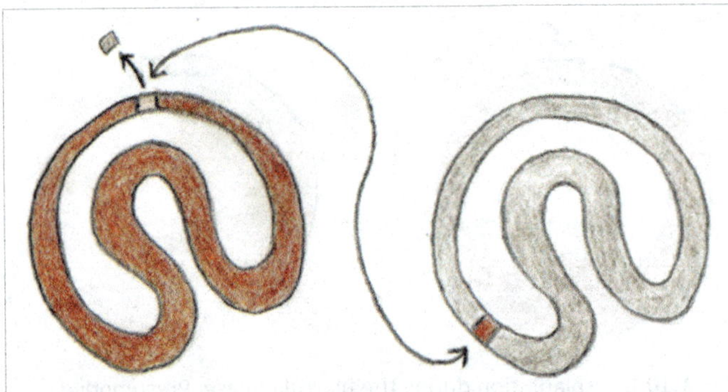

Fig. 1.11 Transplantation during the late gastrula phase. The presumptive nerve tissue does not adapt to its new location, but creates segments of nerve tissue where stomach skin would normally be. The presumptive stomach tissue cannot adapt and is rejected by its new location.

PrO when the newt embryo is further along on its developmental journey. So that is what they did.

Spemann's lab performed the same removal and transplantation operation on newts that were toward the end of the gastrulation process and found that things had changed significantly in the meantime. The presumptive stomach epidermis that was transplanted into the location of presumptive nerve tissue—in that same way as was done in Experiment IV—could no longer adapt to the new conditions and was rejected by the organism at that location. The presumptive nerve tissue transplanted into the stomach skin had also lost the flexibility needed to develop "location-correspondent." Instead, it developed in accordance with its PrO and produced a segment of nerve tissue in its new stomach skin location. It developed "origin-correspondent."

In other words, in both instances the newt tissue had lost the flexibility to adapt to what was called for at its new location. Instead, it remained true to its origins. It was now limited to its original developmental path, its PrP was no longer greater than its PrO. In fact, at this point we can say: the PrP = PrO. In Spemann's terms, the tissue is now determined. No alternative paths are available.

The next step for Spemann's group was to perform similar transplantation experiments across various regions of the embryo. Doing so, they found that at the conclusion of the gastrulation process the tissues throughout the organism had already lost any new developmental potential and were determined, were locked into whatever they would have normally become if they had remained in their location of origin. This was true everywhere, with one exception—and this exception turned out to be an extraordinary one. (Fels et al. 1976; Gilbert 2010; Linder 1998; Nitschmann 1986)

Spemann: Experiment VI

In the process of ascertaining that all areas of the embryo were determined by the end of gastrulation, it was discovered that one particular location was already determined at the outset of gastrulation, well before anywhere else. This location was known as the upper (dorsal) blastopore lip (the UBL), a place that has the PrO to become notochord.

Once it became clear that something out of the ordinary was going on at the UBL, one of Spemann's doctoral students, Helga Mangold, began a series of transplantation experiments there. She was particularly suited for these extremely demanding experiments because of her exceptional manual dexterity. Working with an embryo only 1/16 of an inch in diameter, she would remove sections of the UBL and transplant them into the presumptive stomach area of another miniscule host newt. In order to trace what was happening with the transplanted donor material, it was necessary to use two newt species, one nearly white, the other brownish. Thanks to Mangold's dexterity, the grafted patches not only survived, but—to her and Spemann's amazement—organized the cells around them, causing the stomach area of the host newt to invaginate as normally happens at the location of the UBL, but nowhere else. Not only did the host's stomach invaginate, it continued to develop further throughout the entire gastrulation and neurulation process, resulting in the formation of a second, smaller newt, located face-to-face with the rest of the organism!

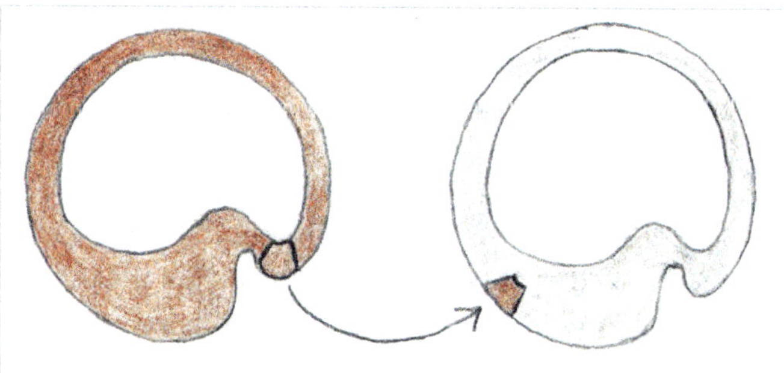

Fig. 1.12 At the outset of gastrulation, a section of the upper blastopore lip (UBL) is transplanted into the area of presumptive stomach tissue in a second newt.

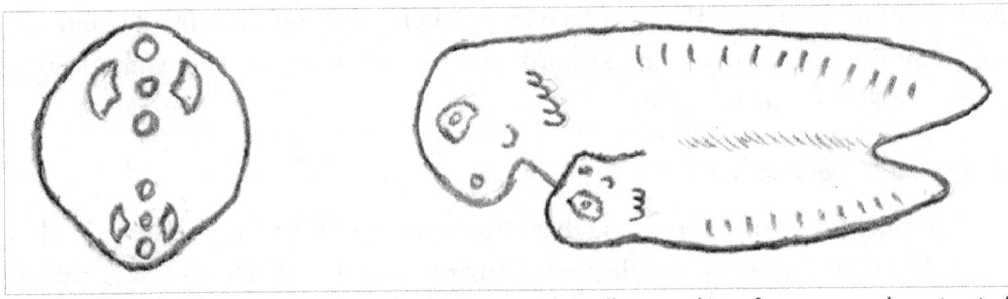

Fig. 1.13 The transplanted section of the UBL organizes the cells around it to form a second newt out of the prospective stomach tissue of the host.

Amazing! The small section of newt tissue located at the UBL would soon enter the annals of modern biology with the unforgettable name: the *Organizer*! In the words of a well-known professor of anatomy at the University of Chicago, Neil Shubin (2009, p. 106):

> *Today, many scientists consider Mangold's work to be the single most important experiment in the history of embryological development.*

The search to determine the components of the organizer-substance led to one of the first scientific research programs that was truly international. Researchers from France, Germany, the United States, Finland, Belgium, Japan and the Soviet Union went to work trying to shed light on the substances at work in the organizer. Yale's R.G. Harrison referred to the amphibian gastrula as the "new Yukon to which eager miners were now rushing to dig for gold around the blastopore" (cited in Twitty 1966, p. 39).

Mangold went on to describe the results of the organizer experiments in her doctoral dissertation, which eventually led to a Nobel Prize in Medicine for Spemann. Tragically, Mangold was unable to share that prize with Spemann. The 11th graders are very saddened to hear the shocking story of how, even before her dissertation had been published, Mangold—at 26 years of age, and the mother of a young child—died of severe burns when a gas heater exploded in her home.

The discovery of the organizer was a huge stimulus for experimental embryologists, and soon further experiments revealed that the organizer was widely distributed in the animal kingdom: It was found in all mammals, birds, amphibians and fish. Much research also focused on better understanding what made the organizer "tick." Its chemical nature was explored with the expectation by many that it would be hormonal in nature. But soon things began to get out of hand—or as

Moore (2003) put it: "out of theory." New experiments subjected organizer tissue to heating, freezing, crushing, and other injuries. Thereafter, it would be implanted into gastrulating embryos whereupon—amazingly—it still induced[27] the development of a second embryo, just as an undamaged organizer would have done.

Other researchers found that even after the UBL tissue had been killed with alcohol, it was still able to induce development. Former Spemann student Johannes Holtfreter, for example, went on to test both living and dead tissues from other animals. In one set of experiments he implanted liver, brain and kidney tissue from mice into the UBL of an amphibian and it still induced the formation of a neural tube.

Such findings amazed researchers and led to the conclusion that the substance of the organizer cannot be the decisive factor, but that residence in the UBL for a period of time somehow enables the tissue to release capacities that are normally dormant (PrP) in it. In many ways the organizer experiments raised as many questions as they answered.[28] As John Moore (professor of biology at the University of California, Irvine) put it in the 1990s: "Half a century later the search still goes on." Nonetheless, the depth of the issues manifesting through such experiments provide the students with a strong sense for organisms as dynamic processes that unfold in time and space in quite amazing and hitherto unimagined ways. (Bonner 1952; Fels et al. 1976; Gilbert 2010; Gray 1967; Nitschmann 1986; Moore 2003; Morange 2020; Shubin 2009; Spemann 1935)

Schotté: Frogs and Salamanders

One related example that brings a nice closure to this aspect of the 11th grade block is an experiment done the first time by Oscar Schotté, a student of Spemann's who later became a professor of biology at Amherst College. It provides a mini-review by bringing together some of the concepts derived from the preceding examples.

Schotté decided to work with the embryos of two animals that belong to different orders of amphibians and have contrasting body structures: frogs and salamanders. Tadpoles have horny jaws, not true teeth, and have suckers on the

27 The term "induction" refers to a process in which an "inducing" tissue calls forth/directs the development of another responding tissue.

28 Although some of the research into these questions—such as the findings in regard to the Nieuwkoop Center—is interesting, in my experience it does not make sense to pursue them further in this context. As one researcher put it, the search for the substances involved in induction was something like "looking for a contact lens in a swimming pool, with the added possibility that the contact lens might be soluble in water." (Morange 2020, p. 86)

sides of their heads. Salamander larva have actual teeth and balancers on the sides of their heads which aid in swimming. With these differences in mind, Schotté took a piece of presumptive flank skin from the underside of a frog embryo and transplanted it into the future head area of a salamander embryo. He also did the opposite and transplanted presumptive flank skin from a salamander embryo in the future head region of a frog. Based on all that they have learned in the past few days, we can ask the class what they would expect to happen now? After some discussion and the formulation of various possible outcomes, the results are revealed.

As it turns out, the transplanted, presumptive frog flank becomes salamander head, but the salamander now has a horny, frog-like jaw (no true teeth), and tadpole suckers at the base of its head! Similarly, the transplanted presumptive salamander flank tissue becomes a head in the frog, but with several key salamander characteristics: true teeth and balancers at the base of the head.

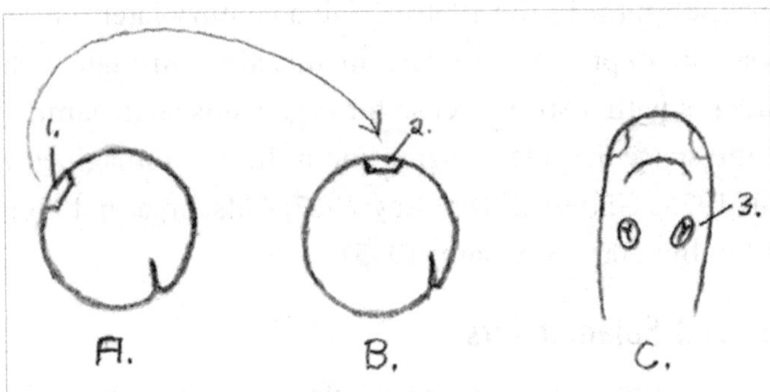

Fig. 1.14 A. Frog gastrula. **B.** Salamander gastrula. Presumptive frog flank skin (1) is transplanted into the presumptive head region of a salamander gastrula (2). **C.** The outcome: the salamander head area now has tadpole suckers (3) and no balancers.

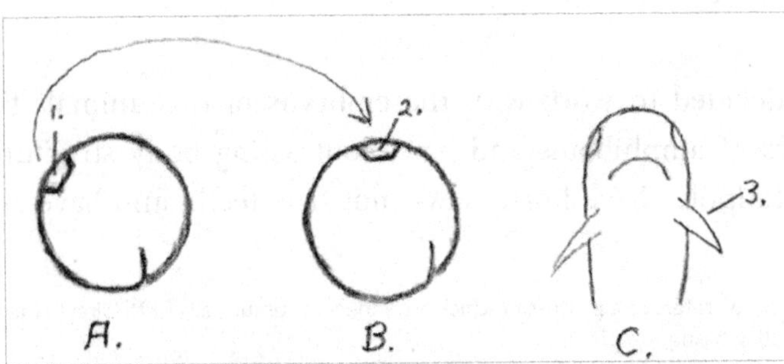

Fig. 1.15 A. Salamander gastrula. **B.** Frog gastrula. Presumptive salamander flank tissue (1) is transplanted into the presumptive head region (2) of a frog gastrula. **C.** The outcome: the frog tadpole now possesses salamander balancers (3) and no suckers.

With this fascinating outcome before us, we can now apply some of the concepts we have recently acquired. I often break the class into small groups and ask them to discuss whether concepts like preformation/epigenesis, mosaic/regulative, origin-correspondent/location-correspondent, prospective-outcome/prospective-potential, and determination shed light on what has taken place here.

Interestingly, we find that by crossing two animals not closely related, categories that are normally "either/or" can also be "and" to some degree. Further discussion with the students lead to conclusions such as the following.

Using the frog flank transplanted into the salamander head region as our example, the overall development is clearly regulative. On the one hand, the outcome is location-correspondent in that the flank tissue becomes head, but on the other hand, it is origin-correspondent in that the head has teeth and suckers, which are frog characteristics that have been carried over into the formation of the salamander head. The same applies to prospective-outcome and prospective potential of the flank tissue. It had the "prospective-potential" to become head, but it stayed with its prospective-outcome to the degree that the head kept certain frog qualities. On the one hand the tissue was determined (it kept its frog qualities), on the other hand, it was not determined since it did not become flank but was able to become head.

After learning about lancelet development and the findings of experimental embryologists, it is interesting to briefly revisit the comparison we made between a lawnmower and a frog at the beginning of the block. Based on what the students have since learned, we can see to what extent the concepts just discussed also apply to lawnmowers and not just amphibians. Such a discussion can grow very lively since the students now realize how organisms are permeated with amazing processes and transformations (mysteries, too) that are far from anything one finds in a common machine/mechanism. (Gray 1967; Hadorn 1961; Moore 2003; Schotte 1938; Spemann 1935)

Wrapping up

As one brings the study of such embryological phenomena to a close it can be enlightening for the students to learn that phenomena similar to those we have been discussing can be also found in the humanities—but on a very different level. The teacher can ask, for example: "What can the study of language (= humanities) tell us, dear 11th graders, that is in sync with the wisdom just revealed to us by developmental embryology?" After they ponder this for a moment, the teacher can provide an example like the following:

If someone comes up to you and says: "You are so hot!" (or cool, or chill), it is essential to know if the person speaking to you is a nurse or doctor at a medical facility, or if it is some dapper regular at the disco. The same statement is not the same statement in different contexts. Words have more than one prospective outcome (meaning). Their prospective-potential is greater than one outcome, because the meaning of a word is significantly shaped by the words around it and the larger context in which they are spoken (written). This was proven for all time—the teacher explains—by the legendary Vince Lombardi, former coach of the Green Bay Packers, who pronounced to the team in no uncertain terms at the halftime of a very important game:

If you are not fired with enthusiasm, you will be fired with enthusiasm!

What he said was said "in no uncertain terms." But as we now know from his statement, and from our developmental embryology studies, it is not the individual terms that make something certain (even in the phrase: "in no uncertain terms"), but the interconnections of the words based on the overall context—on the entire sentence or situation! Such an insight should—has the prospective potential to—fire the students with enthusiasm for the intersection of the humanities with biology!

Microscopy Lab

Studying how organisms come about through continuous division (cleavage) of the egg cell and then following how—through in-folding and out-pouching and the formation of germ layers—the foundation for further differentiation is laid, has hopefully wetted the students appetite for learning how that extremely small world might look like under a microscope.

There are many possible directions a teacher can take at this point. One is to devote considerable time to "magnifying the hidden world" through hands-on microscopic explorations with the students. Finding time to do this during a three or four week block is challenging. In my experience, it is only possible to give a quick orientation into this realm during the morning lesson period itself—the exception being if one has a two-hour session each morning. With that much time, it is possible to devote 30 minutes to microscopy at the end of most periods. If the lessons are shorter than that (they usually are), I think the best approach is to advocate for an extra track class of 45 minutes or so for the duration of the block, or even longer, so that more time can be dedicated to microscopy.

In this way the students can get a genuine taste for this kind of activity, and learn more about its place in the development of modern biology. In this text, I will not go into what one might do in such a hands-on course, for the range of possibilities is great, stretching from the simple experiments found in Richard Headstrom's old classic from 1941, *Adventures with a Microscope*,[29] to college-level experiments outlined in Mary Tyler's lab workbook,[30] that was written to accompany Scott Gilbert's *Developmental Biology.* In any case, a nice way to begin is to compare under the microscope the hair of an adult with a defatted baby-hair (ala Hans Spemann!).

29 Or those found in a book like Kenneth Raines' *Microscope Science Projects and Experiments*.

30 *Developmental Biology: A Guide for Experimental Study* by Mary S. Tyler (2000). This guidebook begins with a discussion of the tools needed to study developing organisms in an experimental fashion and then follows the development of progressively more complex organisms. With the more complex organisms, several experiments are discussed that cover gametogenesis, fertilization, cleavage, organogenesis, regeneration, and how the environment can affect development.

Cytology: The World of the Cell

In the introduction I emphasized that the relationship between the whole and the parts would be a central issue in this block. The work of Spemann and others have made it clear to the students that the whole has a formative influence on the parts both in a spatial and a temporal sense: Location matters, as do the phases of development.

Thus far, we have emphasized the dynamics of such development and have not focused on the details of the emerging "parts" (notochord, epidermis, digestive tract, etc.). In the current context we will presuppose a basic understanding of those organs—which were covered in the 10th grade human biology block—and will now move a step further down the biological ladder into the subunits of the organs: into the cells and their organelles.[31]

So how do we get there? Earlier in the block, the students were surprised that the developing embryo did not increase in size for a considerable time. Instead, through the ongoing process of cell division the cells kept getting smaller and smaller, while the embryo retained its original size. Not only was that a surprise, but it creates a difficulty for those who want to study the organism: The cells soon get so small you can't see them anymore! In fact, due to their smallness it was not known for most of recorded history that living matter is not actually composed of solid substance, but of something quite different: of highly organized and active microscopic units which were given the name "cells."

With the exception of egg cells, most cells are invisible to the naked eye. Typically, less than 10 microns in diameter (= 1/100 mm), they are far smaller than the period that ends this sentence.[32] It was not until microscopes were developed that the abundance of cells was discovered. The students are interested to hear how the name cell came about in 1665 when Robert Hooke used one of the first microscopes (magnification 30X) to examine a thin slice of cork backlit by an oil lamp. He saw that the cork consisted of a huge number of tiny, hollow appearing spaces that were bounded by shared walls. Since this reminded him of the small

31 If we link this to our analogy of the relationship of the sentence as a whole to the words that express its meaning, then we will not be moving here from the overall meaning—the sentence + context—to the words, but from the sentence all the way to the letters that are used in various ways to manifest those words.

32 The teacher can offer extra credit to any student who is able to take a 1 mm strip of paper and cut it into 100 slices in order to demonstrate how wide 10 microns actually is.

quarters that house individual monks in a monastery, he called the tiny spaces cells. About the same time, another early microscopist, Anton von Leeuwenhoek, used an even stronger microscope (almost 300X) of his own construction to discover in a drop of water an amazing world of hitherto unseen single-celled creatures.

But it took another century and a half before biologists fully realized the all-presence of cells. It was not until 1838–1839 that Matthias Schleiden declared that all plants are built up of a multitude of cells and Theodor Schwann[33] made a similar pronouncement about animals. With this, what is now known as the "cell theory" was born (Moore 2003; Raven 2016; Rappoport 2017).

After a brief introduction of this sort—which can be expanded upon if time allows—I have found it effective to move straight into cell division (mitosis), since it provides a good transition from what the students learned about just a few days prior: how cells divide and divide on their way to becoming a blastula, morula, gastrula, etc. We saw this, so to speak, from the outside. The next step is to see in more detail what is going on within the cells with the help of a light microscope,[34] which was the most highly developed magnification instrument when the Spemann/Mangold experiments were taking place.

Mitosis

There are many ways to introduce mitosis to the students, which I will not go into here.

The actual sequence of mitosis can be illustrated by a large drawing on the board. The teacher can then take the class through the phases, naming each and providing labels for the parts. The upside of this approach is the clarity it appears to bring. It has a downside, however, since it accentuates discrete elements and gives the impression that there are distinct phases that must be clearly distinguished, instead of emphasizing the dynamic, flowing process that is actually taking place.

One way to approach this "details vs. dynamic" challenge is to make two run-throughs. The first can provide labels and phase names. The second can emphasize the dramatic appearance and disappearance and of various components for someone viewing mitosis—in process—through a light microscope.

33 Schwann also made the important observation that the cell walls in animals are membrane-like, in contrast to the more robust ones found in plants. He also emphasized the presence of a nucleus as a defining characteristic of cells.

34 In doing this we will be following in the footsteps of Walter Flemming, who, in 1878, was the first to successfully observe the process of cell division.

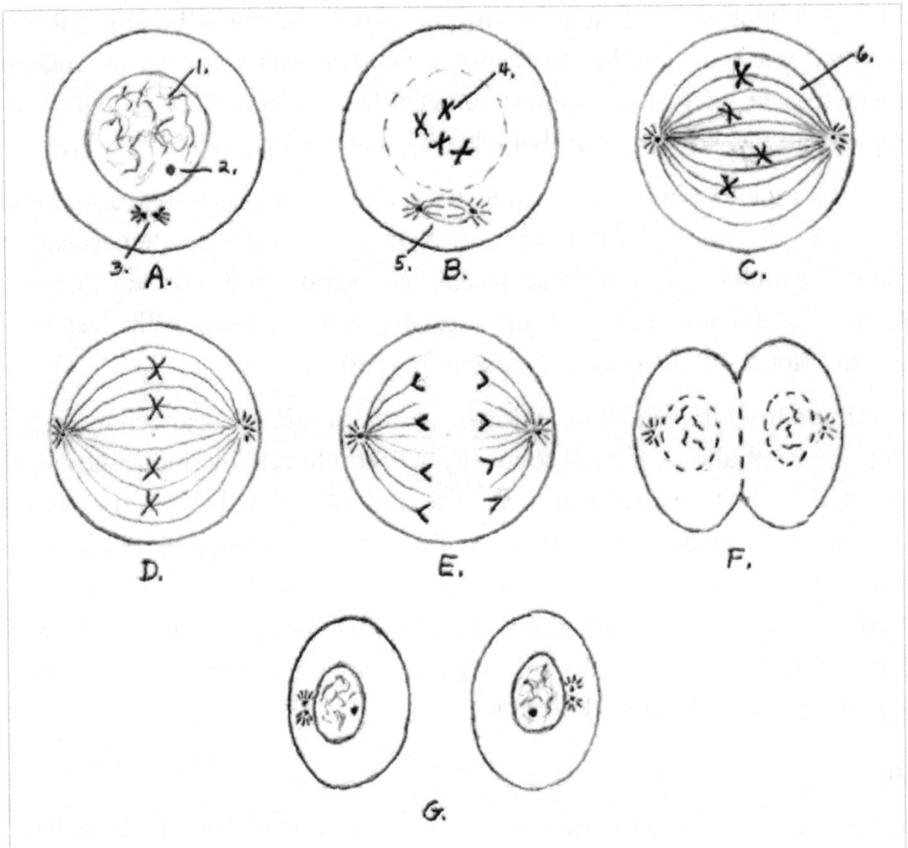

Fig. 1.16 Mitosis: **A.** Interphase. **B.** Prophase. **C.** Prometaphase. **D.** Metaphase. **E.** Anaphase. **F.** Telophase. **G.** Two daughter cells in the interphase. 1. Nucleus. 2. Nucleolus. 3. Pair of centrosomes. 4. Chromosomes. 5. Migrating centrioles. 6. Spindle.

Starting with interphase,[35] a distinct nucleus with nuclear membrane (that houses within it de-condensed—but for the light microscope invisible—chromatin fibers) and a nucleolus that can be observed. Outside the nucleus a pair of centrioles are visible. The cell appears in this form for most of its existence, but then, suddenly, the drama begins!

The nuclear membrane disappears, the nucleolus disappears, and out of what at this level of magnification appears to be "nowhere," something begins to condense and becomes clearly visible: chromosomes with two sister strands (chromatids)

35 It would be possible, of course, to present the conventional G1, S, G2 phases of interphase in the cell cycle, but at the light microscope level nothing can be seen of these processes, so that interphase is best described in a general way as it pertains to growth and to the preparation for later stages.

attached at their center. In the meantime, the two star-like centrioles (asters) have separated and are migrating to opposite poles of the cell. Suddenly they shine like two suns whose rays permeate the whole cell, creating the "spindle." The rays extend and contract repeatedly until certain fibers (microtubules) encounter a connection point at the juncture of the chromatids (the kinetochore) to which they bind and stop their rhythmical shortening and lengthening.

Soon each chromosome pair is attached to two microtubules that come from opposite poles of the cell, and soon thereafter all the chromosomes line up in the center of the cell (the metaphase plate). The quickest and most exciting event during mitosis now takes place: The sister chromatids suddenly separate and are pulled toward the opposite poles of the cell (with their "arms" trailing along behind), while at the same time, the cell begins to elongate, moving the poles further apart.

After the chromosomes cluster at opposite ends of the cell, a nuclear membrane begins forming around each set and the chromosomes start to dissolve into a diffuse network, before disappearing altogether. The spindle also disappears and a nucleolus reappears within each newly formed nucleus. In sync with the forming of the two new nuclei, the cell membrane begins to constrict in a process already seen by the class from the outside: A cleavage furrow appears at the midline of the spindle. As it deepens, it pinches the cell membrane as if lassoed by an invisible noose (not defatted baby hair!) that tightens until the two poles separate and produce two cells out of one.

Looking at this event and realizing that it will be repeating itself again and again, an obvious next question would be: What must be happening in the seemingly quiet phase (interphase) in between? Do we have everything we need at the end (telophase) of mitosis in order to start the process all over again.[36] Discussion soon makes evident that somewhere in the "in between" the single chromosome strands have to become twofold (pairs of chromatids). It soon becomes clear that through the mitosis process each new cell has received the same set of chromosomes. Which leads to the conclusion that the chromosomes must be a very important component of the cell. (Since we will soon be moving through a series of major discoveries in the history of embryology and genetics, we do not jump ahead to the idea of DNA replication at this point. There were many discoveries still to be made before that key insight became possible.)

36 Since some students who are new to the Waldorf School may have learned this already, they can be asked not to respond immediately, so that the others have a few moments to reflect on it first.

What an amazing process! If the teacher is able to present what has just been described with a bit of dramatic flair,[37] it helps the students get a sense for what an incredible event is taking place in this way. In fact, it is estimated that in the human body approximately 330 billion cells are replaced every day in this way!

Once they have a grasp of the overall process, the students benefit from seeing photocopies of mitosis taking place in various organisms. It is also fascinating to see a few quick mitosis videos (ala YouTube) projected onto a screen. And, of course, there exists the possibility to observe this process through a microscope by examining, for example, cells in the tip of an onion root. (Such slides can be prepared with the students or purchased.)

After getting a cytological foothold through mitosis, a logical followup question for the students would be: Can they imagine any cellular circumstances where such chromosome replication (where each new cell gets a full set of chromosomes) might present a problem? This can lead into the question of what must happen when sperm and egg cells conjoin. It is clear, the students realize, that the human zygote, for example, cannot receive a full set of 46 chromosomes both from mom and from dad. This was obvious to researchers soon after mitosis was discovered. They "did the math" (which we can also let the students do) and found that in just 10 generations the number of chromosomes in each human cell would increase to 47,000! So what must happen? Do you get all your chromosomes only from your mother or only from your father? Not likely! So what then? Such warm up questions pave the way for the introduction of reduction division (meiosis), which was discovered in 1883, only a few years after Flemming's discovery of mitosis.

Size Matters.

Going through mitosis and meiosis with large board drawings tends to give the students the feeling that such events are taking place at a level of existence comparable to our everyday life. This is an illusion, of course. It is important to remind the students again of the actual dimensions of the world we are speaking about.[38] As a starting point, we can refer back to the discussions that took place in 10th grade about the significance that size has on our peripheral circulation; how the increase in surface area (SA) to volume (V) frees the microcirculation from

37 In some classes there may even be students who would enjoy acting out the process as a short biological-drama for their classmates.

38 See the interesting book *Why Size Matters* by John Bonner (2006, Princeton University Press) on multiple aspects of this topic.

gravity and allows for "primal flow" in the capillary system. We can add to that how rapidly surface area increases relative to volume when things get smaller. For example, if two cells differ by a factor of ten in radius, the larger cell will have a surface area 100X that of the smaller one, and its volume will be 1000X greater! (For spherical cells, the SA is proportional to the square of the radius, whereas the V is proportional to its cube.)

Why does this matter? Because the SA is where the cell interacts with its environment; the surface is where all substances enter and exit the cell. When cells get larger, as some neurons do, they manage this by getting long and thin, so that at all points the cell membrane is close by and diffusion between inside and outside is easily realized.

The students learned earlier that a typical animal cell is around 10 microns (μm) in diameter (= 1/100mm). So how does this relate to a few things the students are familiar with?

- A fog or cloud droplet is about 10 μm in diameter.
- The width of a strand of spider web silk is 3–8 μm.
- Paper thickness ranges from 70–180 μm. (So if you slice a piece of 100 μm thin paper into 10 slices, you get a typical cell diameter.)
- A typical cell is usually about 1/5 the size of the smallest particle visible to the naked eye.

Another way to give students a sense for how small the world of cells is, is to describe all the effort that goes into making microscopic observations possible. (Alberts et al. 2007; Rappoport 2017; Raven et al. 2016).

Cellular Microscopy: What It Takes

Animal cells are not only hard to observe because they are so tiny, but also because they are colorless and translucent. The challenge of seeing the detailed features of a living cell has been compared to the difficulty of seeing the internal characteristics of a clear glass marble that is immersed in a transparent tumbler of water.

We explain to the students the various techniques that scientists developed to overcome this challenge. The first was to apply stains that colored the cell's components, which made differences in structure visible. But before staining, a permanent preparation had to be created by treating the cells with a fixative that immobilized, killed and preserved them. Since most tissue samples were too thick for examining individual cells at high resolution, the tissues had to be cut into very thin slices with a tool (microtome) that operates like a meat slicer. These sections could then be laid flat on a glass microscope slide. However, since tissues are usually soft and fragile even after being fixed, they needed to be embedded in a supporting medium—usually waxes or resins. These could then be hardened to form a solid block that could be easily sliced by the microtome.

However, despite all of these innovative techniques and more than a century and a half of experience, there is still a serious possibility that fixation and embedding may alter the structure of the cell and its substructures, thereby creating what are known as "artifacts." An alternative method is rapid freezing, which eliminates the need for fixation and embedding since the frozen tissue can be cut directly. In this way, certain artifacts can be avoided. However, with this method the fine structure of the cell is often disturbed and displaced by ice crystals. Another difficulty caused by slicing tissues into thin sections (the thinner the section, the crisper the image) is the loss of the third dimension. This challenge is addressed by taking a series of optical sections at different depths, which are then reconstructed into a three-dimensional image with the help of a computer.

After the sections have been cut, the next step is staining. A very popular approach today is to stain cells with fluorescent dyes, which are then viewed under a fluorescent microscope. Unfortunately, fluorescence microscopy is not completely noninvasive, since the high-intensity excitation light is inherently toxic for live cells. Physiological changes induced by such excessive illumination can lead to artifacts and atypical responses.

More could be said about such challenges and the approaches developed to overcome them (such as the confocal microscope), but the important thing for the students to realize is that what they will soon be learning about cells and their subunits is not simply based on enlarged photos of a miniscule world.

In many ways this is even more true for what is revealed by electron microscopy. Although theoretically the resolution of an electron microscope would be 10,000X that of a light microscope, due to various hindrances their resolution is only about 100X greater (which is still significant!). New procedures for embedding, cutting and staining tissues had to be developed for the electron microscope. Because electrons have very limited penetrating power, the fixed tissues have to be cut into extremely thin sections—about 1/200 the thickness of a single cell. (The students can try to imagine how thin that is by reminding themselves that you would get the size of a cell 10 μm in diameter (= 1/100mm) by slicing a 1mm section of paper into 100 slices. Now, to get the thickness of the cell section put under an electron microscope you would have to slice that slice into 200 slices. Such dimensions are not part of the world we consciously live in! Such slicing is done with a diamond-bladed microtome (metal microtome blades are too soft and dull to cut ultrathin sections) after the cell has been dehydrated and permeated with resins that form a solid block of plastic.

Again, researchers must ask: How can we be sure that the image of the dehydrated, resin-embedded and fixed specimen bears a strong relationship to the very delicate aqueous system that was present in the original cell? As with light microscopes, rapid freezing is one approach currently used to mitigate some of these difficulties.[39]

Since protein molecules, for example, can only tolerate electron doses in orders of magnitude below what is needed to create a defined image, tens of thousands of individual images must often be combined (with the help of a computer) to produce an average image that reveals details hidden by the noise in the original images.

Determining the detailed structures of membranes and organelles in cells requires high resolution images. Multiple images and views can be combined from different directions to produce detailed reconstructions using a technique known as EM-tomography. A scanning electron microscope (SEM) is a simpler device that is able to produce an image of the three-dimensional structure of a surface of a specimen. But first it has to be fixed, dried, and coated with a thin layer of heavy metal—or alternatively—rapidly frozen.

Why all this detail? As mentioned above, it is meant to help the students realize how far we are from the everyday sense world when learning about cytological details. We are not only far away, but a huge amount of technological savvy has been necessary to put together the representations of cells that are presented to us. (More on the limitations such cell-representations shortly.)

It is also interesting to note that according to the widely regarded historian of molecular biology, Michel Morange, apart from two key discoveries (the structure of the bacteriophage and the existence of ribosomes) the electron microscope has played only a secondary role in the growth of molecular biology, even when it comes to the identification of intracellular structures.

39 Very recently, new technologies—often referred to as "nanoscopes"—have led to resolutions greater than what had previously thought to be possible.

> *The images produced by the electron microscope were in fact too complex; to correctly interpret the images, scientists needed some kind of preexisting model of what they were looking at. Furthermore, the energy from the electron beam was so great and the chemical treatments involved in stabilizing the sample and increasing the contrast were so extreme that the technique produced a large number of artifacts and made pseudo structures appear where none were present. Images produced by electron microscopy no doubt had a certain pedagogic value and were useful as documentary proof, but they led to few if any major discoveries. Instead, electron microscopy simply confirmed the reality of results that had generally been obtained using other techniques.* (Morange 2020, p. 98; Alberts et al. 2007; Baer et al. 1971; Magidson & Khodjakov 2013; Rappoport 2017; Raven 2016)

Cells and Organelles

What, then, do we find after all the effort that has gone into developing instruments and techniques for "seeing" into a world that for humans is invisible? One challenge starting out is not to get lost in taxonomical distinctions. A simple way to do this to provide a few taxonomical categories, such as the difference between eucaryotic and prokaryotic cells, and then to highlight a few key differences between plant and animal cells, such as cell walls, central vacuoles, and chloroplasts. Due to time constraints and all the exciting material still to be covered in this block, I then put the focus on animal cells—which have been our topic up to this point, anyway.

What do the enormous microscopic efforts just described above reveal to us about the nature of animal cells?

The substructures found in animal cells can be introduced to the students with the help of a large generalized chalkboard drawing such as the one below (Fig. 1.17).[40] In my experience, if the students are not to "die on the vine" during the introduction of these organelles, then only a brief characterization of each major cell component (plasma membrane, cytosol, endoplasmic reticulum, etc.) should be given. I do not expand upon the role of nucleus yet, since the dramatic discovery of DNA's significance is a major part of what is still to come in this course.

40 In order to do justice to the considerations just made—that this is not a tiny version of the world we know—I think it important *not* to make this a colorful, three-dimensional image such as one frequently sees in textbooks. Gray in gray makes it is easier to communicate that this is a reconstruction of a great many images compiled in ways just discussed.

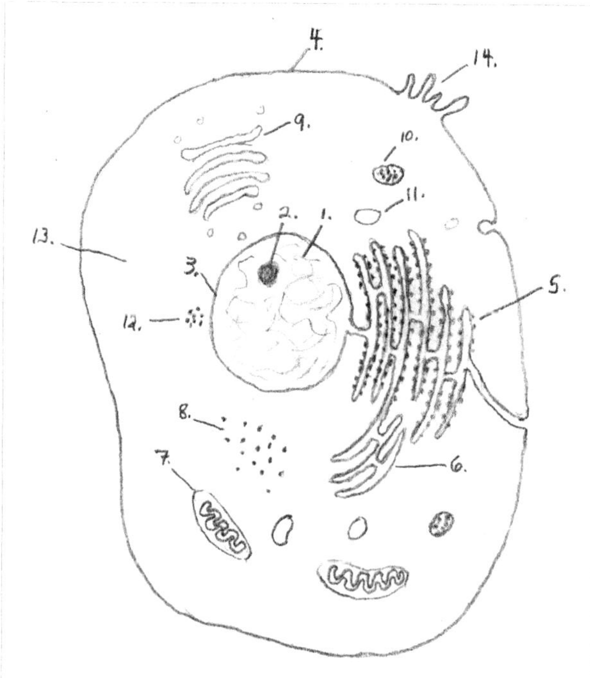

Fig. 1.17 Static representation of major cell components reconstructed from electron micrograph data. 1. Nucleus. 2. Nucleolus. 3. Nuclear membrane. 4. Cell membrane. 5. Rough endoplasmic reticulum. 6. Smooth endoplasmic reticulum 7. Mitochondrion. 8. Free ribosomes. 9. Golgi complex. 10. Lysosome. 11. Vacuole. 12. Centriole. 13. Cytoplasm. 14. Microvilli.

Such a cursory coverage of the organelles and cytoskeleton stems from a perspective that has been repeatedly emphasized in the 9th, 10th and 11th grade biology blocks. It involves the tension between the exhausting effect that comes from the explication of too many details, on the one hand, and the importance of conveying the process-nature of living organisms, on the other. When teaching such topics, it is an ongoing challenge to find the right balance between the extremes of a) dying in the details and b) getting lost in the neverending transmogrifications that are taking place.

This tension takes on a very interesting form at the ultrastructure-level of the cell, since—it turns out—to represent the cell as made up of permanent structures, such as they appear in the diagram above, is actually misleading.[41] Well known

41 The topic of how the pictorial representations of cells in textbooks today actually relate to the reality they are referring to is an important one. The nature of the representations in molecular biology has been the focus of

continued p.56

Austrian biologist Paul Weiss gives a clear explication of this by comparing a cell to a factory:

> *Taken by itself, a static picture (of a cell) fails to reveal whether it portrays a momentary state of an ongoing process, or a permanent terminal condition. Unless this ambiguity is constantly borne in mind, one runs the risk of mistaking the static picture of the cell for evidence of a mosaic of well-consolidated structures.... A cell works like a big industry, which manufactures different products at different sites, ships them around to assembly plants, where they are combined into half-finished or finished products to be eventually, with or without storage in intermediate facilities, either used up in the household of that particular cell, or else extruded for export to other cells or as waste disposal... There is a major flaw, however, in the analogy of a cell and a manmade factory. While in the latter, both building and machinery are permanent fixtures, established once and for all, many of the corresponding subunits of the cell (organelles) are of ephemeral existence, in the sense that they are continually or periodically disassembled and rebuilt, yet always each according to its kind and standard pattern. In contrast to a machine, the cell interior is heaving and churning all the time.... The only thing that remains predictable amidst the erratic stirring of the cytoplasm and its substructure is the overall pattern of dynamics.* (Weiss 1971, pp. 20–21)

In a similar manner, another expert in the field, biochemist and molecular biologist Franklin Harold, describes how the cytoskeleton "is subject to frequent remodeling. Mitosis, for instance, entails the dissolution of much of the cytoskeleton; its components are redeployed in the service of cell division, and subsequently reconstituted in their former order. Everything is in flux, but in a regulated purposeful manner." He speaks of "the dynamic nature of the endomembrane

---

a number of investigations. Michel Morange (2020), for example, devotes an entire chapter at the conclusion of his 500-page history of modern cell biology to "Images, Representations and Metaphors in Molecular Biology." Norberto Serpente's (2011) explication of this issue is well worth reading. He concludes:

> *Over centuries cells have been the target of optical and electronic microscopes as well as other technologies, with distinctive types of visual output. Whilst optical technologies produce images "evident to the eye," the electronic and especially the molecular create images that are more elusive to conceptualization and assessment.... Electron microscopic images are considered to be instances of "seeing the object" more closely. In my opinion, this position is debatable because electron microscope images have an ambivalent nature. For the electron micrographs, rather than being just instances where our sight gets close to the cells, are instead a composite of nature's configuration and of mental modeling. Therefore, electron microscopic images are more difficult to conceptualize as images resembling the object due to the intricacies involved in their production and to the fact that they lack a clear referent to compare with.*

system. What we refer to as the ultrastructure is more like a snapshot of a fluctuating flow of membranes and substances associated with them, from the nucleus to the periphery and from the surface to the center." (Harold 2001, p. 122, 124)

As it turns out, membranes at the ultrastructure level reveal an amazing convertibility from one structure to the next. Depending on the momentary process under way, the cell membrane

- can bud off (invaginate), creating a vessel for fluids (pinocytosis);
- can become a vessel for transport through the cell (cytopempsis);
- can become an enveloping membrane for the digestion of substances (phagocytosis);
- can reunite and become cell membrane once more as a content is sent forth from the cell (exocytosis);
- can build E-R.

E-R, on the other hand, can be used to build the cell membrane, or nuclear membrane. Golgi vessels, after transporting a substance can also become cell membrane. And so on.

All of these transitory structures are determined by the momentary metabolic processes taking place and are thus very different from the spatial and form consistency we find at the macro-level of the organism. "Form follows function," and since function is constantly changing in conjunction with the needs of the organism as a whole, so is the cell's ultrastructure.

So what is the source of all the energy that enables this constant heaving and churning? Although it was probably mentioned briefly when the teacher gave an overview of the cell organelles, the above question provides a good opportunity to delineate briefly the key role that mitochondria play in the ATP-ADP cycle.[42] The fact that some cells can have thousands of mitochondria and others only one should become clear out of that discussion.

So if different cells are doing very different things at any moment in time, what determines what is going on in a particular cell at a particular time? Although the answer should be very obvious, the cell-centered focus we have been practicing the past few days can result in "forgetting of the forest" and lead to an to an explanation of everything from the limited perspective of the individual trees. It is thus important to underscore in discussion with the students that the individual cells

42 More detailed coverage of glycolysis and the Krebs cycle can be left to the biochemistry block.

are all subunits of tissues, which are parts of organs, which are active in the context of the whole organism, which itself must be seen in relationship to its environment.

Such a discussion can also provide the transition to another central cytological fact: Not all cells are alike. All students know that nerve cells must be different from liver cells, from kidney cells, and so on, which leads to the fascinating question: How did the different cells become so different when they all started out as subdivisions (blastomeres) of the same zygote? (Harold 2001; Rappoport 2017; Schneider 1964; von Kraft 1968)

Cell Differentiation

Up until now we have just looked at what eukaryotic cells have in common. Earlier in the block we noticed that cells at the animal pole became smaller than those at the vegetal pole. We learned, too, about ectoderm, mesoderm and endoderm as a first differentiation that would lead to different organ systems. And we spoke about how cells/tissues had a prospective outcome, but also a greater "prospective potential," at least for a while. With our current focus on cells, we can now ask: How does all this play out on a cellular level? A large diagram on the board is the key to the kingdom here (Fig. 1.18).

Although the cell shapes at the bottom of the diagram are more symbolic than literal, they give a sense for the diversity of specialized cells that exist. The differentiated and labeled cells at the bottom clearly fall into three major groups. The students will be reminded of our embryonic studies a few days prior and some will recognize—before the teacher has labelled them—that the three large groupings must be ectoderm, mesoderm and endoderm. As we move downward, we see how the range of potential narrows until specialized cells mark the ends of each developmental pathway. The cells above the lowest level are termed "multipotent" because they can produce more than one kind of specialized cell, whereas those that bring forth the multipotent germ layers are termed "pluripotent." Only the zygote and its first few subdivisions are considered "totipotent," meaning they can bring forth an entire organism and all its diverse cell types.

After we have thoroughly discussed the diagram and what it represents, we can use it to reflect on a number on interesting topics. (It provides, namely, a great opportunity for what Rudolf Steiner calls an "out-breath," where we take a narrow topic and expand it out into wider perspectives.) To start off, it is interesting to compare the cells at the top and the bottom of the diagram. The upper cells have the potential to develop in many directions, but undeveloped as they still are, they

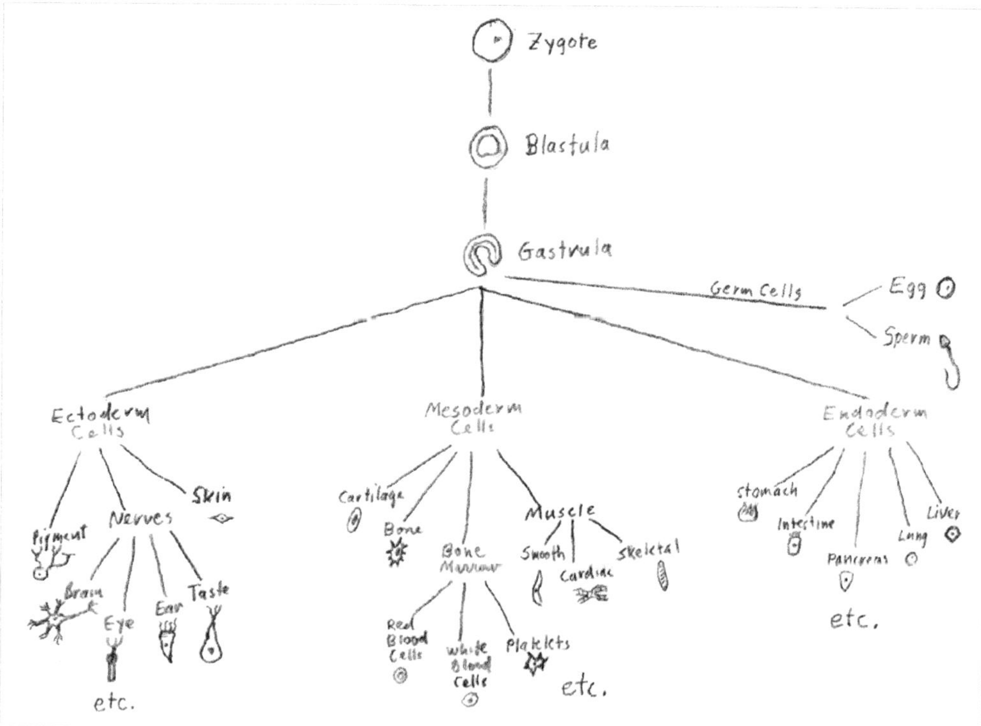

Fig. 1.18 Simplified diagram of cell differentiation, which in human beings produces some 200 different cell types.

can "do" nothing—they are not yet functional. At the bottom, by contrast, we have a whole range of very defined, one-sided "capacities," functions that each cell-type performs within the larger whole. Since these functions are very limited in scope, however, they must be interwoven with many others.

From here it is possible to "breathe out" into a wider perspective by asking, for example: Can we see any similarities here to the course of a human biography? In discussion it becomes clear how open we are as young children—we can learn any language, for example—and are not fixed in the path our lives will take. But what happens as we grow older, does our path narrow? What role do predispositions, what roles do our environment and our education play in what we become? Are we totally fixed age in the way that a specialized cell is when we reach old age: totally one-sided and unable to change? Or do we still have some flexibility? Do we as humans have a prospective outcome that we will eventually be locked into, or do we retain some prospective potential throughout life, if we choose to tap into it? Many questions of this sort can be considered.

Another interesting question to consider is how such things are in the animal kingdom: What is the difference between a young lion and an old one, and what happened in between? Do they follow a trajectory similar to the human one when moving from childhood to adulthood? What are the similarities, what are the differences?

Regeneration

After learning about the path to differentiation that cells go through in a developing organism, it is interesting to determine the extent to which an organism maintains this capacity once it has reached maturity. One avenue for pursuing this question is to study how organisms react to the loss of body parts. A leader in this field was the Swiss zoologist Ernst Hadorn. One of his pioneering studies involved amputating a salamander's leg and recording in detail how the organism dealt with that injury.

The developmental process can be illustrated with five or six chalkboard drawings that highlight the central transformative phases.

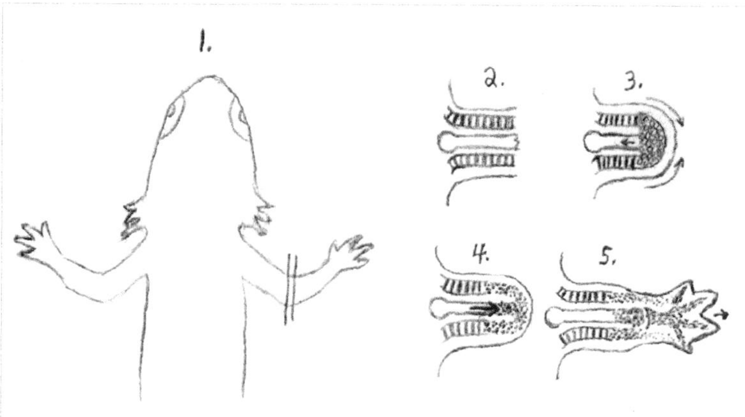

Fig. 1.19 Salamander leg regeneration. 1. Double lines show the amputation location. 2. Post amputation with muscle (striped) above and below the severed leg bone. 3. Epidermis grows over the wound. Bone and muscle cells break off into individual cells and lose their differentiation. 4. Cells begin to reform the limb pattern. 5. Cells begin to grow, re-differentiate—even transdifferentiate—and re-form the entire limb structure.

1. After the limb has been amputated just above (proximal to) the elbow, the epidermis begins to grow and within one day has covered over the wound, forming an apical epidermal cap.
2. Inside the closed wound the bone, muscle, nerves, etc. begins to dissolve and the cells separated off into individual cells. This process progresses inward from the stump until a considerable mass of cells has been created. As this is happening, the cells also begin "de-differentiating"

and taking on the appearance of undifferentiated embryonic cells. The cell mass that results is called the "regeneration blastema."

3. The cells of the regeneration blastema begin to divide, grow outward and reform the structure of the limb.
4. As this happens, the cells begin to "re-differentiate" into bone, muscle, cartilage, etc. in the appropriate places, while also forming joints and digits, until a normal limb is recreated. Highly interesting, too, is that not all cells return to their original type. Some cells "trans-differentiate." For example, cells that were previously differentiated into muscle cells might now become cartilage tissue! After about six weeks, the entire limb pattern has been restored through this process, and with some added growth becomes almost undistinguishable from the original.

A process such as this—whether manifesting embryonically or as regeneration—is often called "morphogenesis," which refers to the generation and coordination of form and structure during development. Amazingly, the class now realizes, animals such as salamanders still retain the morphogenetic capacity to reproduce a significant body part all the way into adulthood.[43]

After reviewing the recent literature about this theme, Peter Heusser (2016, pp. 121–122) concludes:

> *The more detailed the present-day findings about morphogenesis become, the clearer it is that morphogenesis only takes place thanks to a precise dynamic orchestration of all the process-cascades involved....* [These are] *always mutually dependent processes which are coordinated spatially and temporally with other processes... an active coordination of all the sub processes in the sense of a unity in the whole.*

Although the kind of regeneration revealed through Hadorn's experiments goes far beyond anything available to injured humans (with the exception of the

43 In addition to cell differentiation, de-differentiation, re-differentiation and trans-differentiation, there is even more "cell-shaping" going on as the organism morphogenetically creates the intricate forms of its tissues and organs. For example cells often migrate to reach their appropriate locations; groups of cells can come together and create various shapes (roll together to form a tube, for example); a cell's activity can cause significant size changes (for example: egg cells conserve and add cytoplasm to become "huge," while sperm cells eliminate most of their cytoplasm, causing them to grow smaller). Beyond such cell-shaping activity, the regulation of cell division referred to at the outset of this block is fundamental to morphogenesis, as is regulated cell death (apoptosis), which we will soon take up in more detail. Morphogenetic processes such as these are yet another way in which "the parts" are shaped and organized in service of "the whole." (Gilbert 2010, Maienschein 2007)

regeneration of liver tissue[44]), it raises interesting questions about why different creatures have this capacity to such varying degrees? Most students know how lizards can lose their tails and regrow them. This is, however, not a complete regeneration like we found in the Salamander. The new tail no longer has a vertebral column, but instead a cartilage rod that serves as its axis. Moving up the phylogenic ladder to birds and mammals, even this form of regeneration is no longer possible. What we find at this level is the healing of wounds—a process where the cells in the body begin dividing more rapidly in order to repair a wound. Scar tissue often forms in conjunction with this process. Healing of this nature stands in contrast to the kind of regeneration observed in Hadorn's experiments, which involved the actual rebuilding (morphogenesis) of a damaged/missing body part.

If we look in the invertebrate direction, however, a notable capacity for regeneration is evident. Here are a few examples:

- Regeneration in arthropods is hindered by their hardened exoskeleton, so if they lose a limb, they must wait until the next molting, when the lost limb appears in reduced form. It will, however, continue to grow further with each subsequent molting.
- Sea stars can regenerate lost arms, even if all of them are lost. Some species can regenerate an entire sea star from a detached arm if it contains at least 1/5 of the central disk.
- Slugs and snails can re-grow their eye stalks.
- The students are usually familiar with earthworms and know that the head region can re-grow a new tail (if cut behind the clitellum), but that the tail cannot re-grow a head. They often do not know, however, that under the right conditions flatworms (*planaria*) can be cut into some 200 pieces and within a few weeks each piece is able to regenerate a complete organism.
- One species of marine polyp (*Hydractinia*) known as "snail fur," frequently has its head bitten off by grazing fish, and can re-grow it within a few days![45]

44 Which, however, regenerates without cells dedifferentiating and forming a regeneration blastema, but through the increased division of already differentiated cells. This is known as "compensation regeneration."

45 Obviously, the "head of a snail fur" is not a highly differentiated organ that performs all the functions we normally associate with heads.

- Sponge cells can be put through a fine sieve so that the tissue structure is destroyed and the cells separated. If allowed, the cells will aggregate again into groups and whole new sponges can develop from them.

Reflecting on all of this, the students recognize that the animals with the most regenerative capacities are those that appear most plant-like to us. (In fact, many students think that a sponge, hydra, or sea anemone is an exotic marine plant when they see one for the first time.)

A defining feature of animals as we move up the phylogenic tree to birds and mammals is the highly refined Central Nervous System (CNS). The development of more wakeful consciousness with the help of the CNS is, however, accompanied by a loss in vitality. This is evidenced by the fact that "the CNS is the only tissue in the body that does not regenerate" (Illis 2012, p. 259). As mentioned above, other organs can heal through increasing the rate of cell division, but not the cells of the CNS, which stop dividing soon after birth.[46] (Neural plasticity, discussed in the 10th grade, is not attained by cell division, but by increasing connections between existing neurons.)

Without realizing it, the students were already introduced to the polarity between vitality and consciousness at the outset of this block. Early in the development of the egg cell, we observed that the cell cleavage-plane was not located directly at the cell equator, but somewhat above it. This created larger, yolk-accentuating cells in the lower part of the egg and smaller ones toward the top. Scientists have given these diverging tendencies names: The lower part of the egg is known as the "vegetal pole" (the "vitality pole" in German) because it is the basis for the vitality-producing digestive organs. The upper part is called the "animal pole" (animus in Latin = soul or mind) because through neurulation it becomes the bodily basis for consciousness—the nervous system.

Anyone who is forced to stay awake for extensive periods of time knows that consciousness and vitality are contrasting principles. As we all know, we can only sustain our bodily vitality by shutting down our consciousness on a regular basis—a condition known as sleep—by means of which we re-enliven ourselves mentally and physically. (Gilbert 2010; Gurtner et al. 2008; Hadorn 1961; Hickman 2001; Husemann 1974; Logan 2014; Okada 1995; Reddien & Sanchez 2004; Tsonis 1995; Tsonis & Fox 2009)

46 Although reports vary depending on the source, it appears that the hippocampus is still able to produce several hundred new neurons per day during childhood (till around the age of 13 some studies report), which is a very small number considering that the human brain consists of some 100 billion neurons.

Cells in Vitro - Cells in Vivo

Cell differentiation and various forms of morphogenesis have made it evident to the students that cells within an organism are subordinated to a "dynamic orchestration" of "mutually dependent processes, which are coordinated spatially and temporally with other processes..." (Heussler 2016). An interesting question that we can now put before them is: How do they think cells would behave if we were able to remove them from the organism, to remove them from the living context in which they play a specific role?

Over a seven-year period, Leonard Hayflick and Paul Moorhead performed a pioneering set of experiments that shed some light on this at the University of Pennsylvania. Hayflick and Moorhead removed fibroblast tissue[47] located in the lungs of a four-month-old aborted human embryo and broke it down with digestive enzymes into separate cells. These cells were then cultivated in vitro[48] by incubating them in glass bottles together with a suitable growing medium. Within a few days, the cells spread out on the glass surface and began to divide. They kept dividing (doubling) over a time span of about seven months. During this time period, the fibroblast cells doubled approximately 50 times, which—if all the cells had been kept active in the experiment[49] —would have generated some 20 million tons of cells. Taken away from the constraints and formative influence of the organism, cells that would have become part of an infant's lung tissue became instead an unimaginable mass of cells—20 million tons worth!

As studies of this type became more frequent, it was found that normal cells cultured in vitro frequently undergo spontaneous transformations, after which they are able to divide and multiply indefinitely when they receive the appropriate nutrients and sufficient space. They are then commonly referred to as "immortal cell lines." Often used for experimental purposes, thousands of such cell lines have been established in vitro from hundreds of different species. (Hayflick 1968; Kaur & Dufour 2012; Waddington 1968)

47 Fibroblast tissue builds fibers in connective tissue

48 Which means, as we explain to the students, "in glass," as opposed to "in vivo," which means "in the organism."

49 After only 10 doublings the cells would have already covered the surface of 1024 bottles. Therefore the researchers had to freeze and/or eliminate many of them, and could only project the final numbers based on those that were maintained.

HeLa Cells

If time allows, the students are usually fascinated—and shocked—by the story surrounding the most famous cell line of all, the HeLa Cells. They became known as HeLa Cells in reference to the name of the young woman from whom they were taken: Henrietta Lachs. In February 1951, Henrietta visited Johns Hopkins Hospital because of a painful knot in her cervix. That same day she was diagnosed with cervical cancer. Prior to treatment, cells from the carcinoma were removed for further research purposes without Henrietta's knowledge or permission—which was standard procedure at that time. Despite various treatments, the cancer spread throughout her body and Henrietta Lachs died 9 months later at the age of 31. Her story, the struggles of her family, and the many ethical issues that arose around the use of her cells (that came to the fore decades later) cannot be expanded upon here, but have been thoroughly documented in the widely read book, *The Immortal Life of Henrietta Lacks* (2010) by Rebecca Skloot.

From a purely biological perspective, the HeLa cells were an essential laboratory tool in many significant biomedical breakthroughs. The cancer cells that were taken from her body before she died have never stopped dividing and multiplying.[50] They are being employed for experimental research in laboratories around the world. They were used, for example, by Jonas Salk to develop the polio vaccine, and by AIDS researchers in the 1980s to identify and isolate the human immunodeficiency virus (HIV). They were even launched into space for experiments in zero gravity. Research that involved HeLa cells has also served as the underpinning of several Nobel Prize winning discoveries. Some 70,000 studies have been published involving the use of HeLa cells, and over 110,000 publications have cited their use between1953 and 2018. And, to return to our topic of "immortal cell lines," Rebecca Skloot notes in the aforementioned book: "One scientist estimates that if you could pile all HeLa cells ever grown onto a scale, they'd weigh more than 50 million metric tons—an inconceivable number, given that an individual cell weighs almost nothing (p. 2)."

Interesting reflections on the nature of organisms can follow from such amazing findings. For we are seeing what happens to cells when they are taken out of their living context, abstracted, so-to-speak, from the web of interactive relationships that are the signature of an organism. Within the organism, the life

50 The downside of their exuberant growth is that they have contaminated many other cells lines and invalidated significant research projects over the years.

of cells is tightly regulated.[51] As has often been mentioned here, "if all the cells in our legs were to divide just one more time...": An NBA[52] career would almost be a given! In our 10th grade block we discussed examples of cell regulation: how when we go up into the mountains the production of red blood cells immediately begins to increase, when we go back down again, more red blood cells die than divide. Everyday life teaches us that if you cut yourself, your skin cells will divide rapidly to close the wound and then stop, and so on. Throughout the organism is an ongoing regulation of cell production and cell death that is based on the roles those cells play within the greater whole. (Grady 2010; nih.gov 2020; Schmeck & Schmeck 1986; Skloot 2010)

Apoptosis

Another interesting aspect of organized cell activity in vivo is the regulation of cell death (apoptosis), which plays an important role in an organism's normal growth and development. The embryonic development of human hands and feet provide a classic example of apoptosis—one that leaves the students with a vivid image of how necessary organized cell death is for the functioning and shaping of an organism.

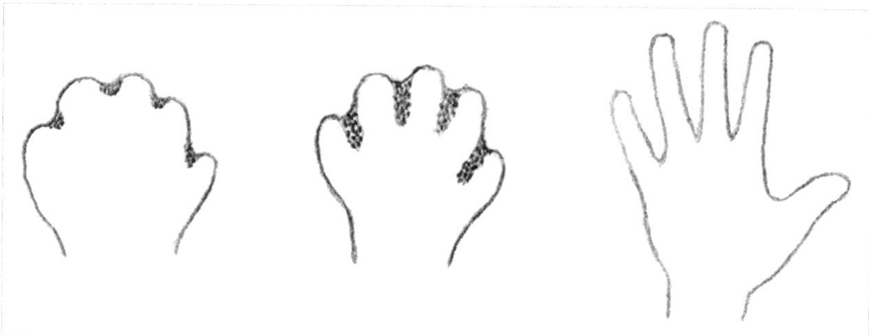

Fig. 1.20 Sculpting of the human hand through regulated cell death (apoptosis) in the interdigital spaces.

For starters, we can let the students speculate on how the form of their fingers might have come about embryonically. As it turns out, it does not happen—as they often imagine—because certain cells divide rapidly in a linear fashion creating long digits, but because the cells between the growing finger tissue die away (apoptose), creating the separate fingers. If they hadn't done so, we would all have webbed hands and feet, reminiscent of what we find in many semiaquatic creatures.

51 The exception to this rule are cancer cells, which will be addressed shortly.
52 aka professional basketball

Apoptosis involves the death of cells, but in contrast to what people tend to think about death processes, this kind of death benefits the organism as a whole. Not only does apoptosis work as a sculptor of many organs, but it prunes unneeded structures and controls the number of cells in certain tissues. Either too much or too little apoptosis is a factor in many human ailments, from degeneration/atrophy (too much death) to cancer (too little death). In adult humans, as many as 100 billion cells die in an ordered fashion each day and are replaced.

The way in which cells die and are eliminated during apoptosis is interesting to compare with cell death caused by harmful factors such as toxic chemicals or injury (necrosis).

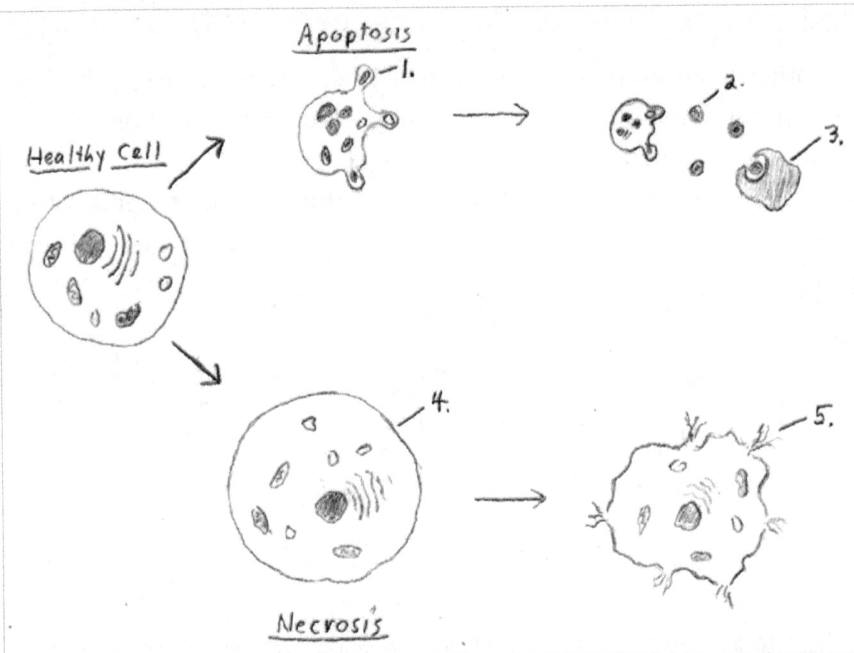

Fig. 1.21 Apoptosis in contrast to Necrosis. 1. Bleb. 2.Apoptotic body. 3. Phagocytosis of apoptotic cell. 4. Cell swelling. 5. Membrane rupture and leakage of cellular content.

In the latter instance, the damaged cell's plasma membrane can no longer control the passage of water and ions, so the cell swells up, and its contents leak out through the membrane. This often causes inflammation in the surrounding tissue. Cells undergoing apoptosis, by contrast, go through a much more orderly process. They begin to shrink and the plasma membrane forms bubble-like protrusions called "blebs" that form around different organelles. These blebs then separate from the cell and form "apoptotic bodies" that attract phagocytes (macrophages), which consume them, thus preventing cytoplasmic leakage into the interstitium and inflammatory reactions in the surrounding tissue. (Elmore 2007; Gilbert 2010; Marieb & Hoehn 2012; Morange 2020; Wickman et al. 2013))

Cancer

We can follow the discussion of such processes with a question to the students: What would happen if the opposite of apoptosis—uncontrolled cell proliferation that "ignores" the organization of the larger whole—were to take place in an organism? In other words, what if the tendency that cells show outside the organism in vitro—to divide and divide and divide—were to manifest in the body? Usually several students will realize that something like that take place with cancer. Cancer researcher and professor of medicine at Columbia University, Siddhartha Mukherjee, puts it like this: Cancer "is a testament to the unfathomable power of cell growth." Uncontrolled cell division no longer heeds the morphogenetic forces discussed earlier in this block. What is normally regulated to perfection—cell division and controlled cell death (apoptosis)—takes on its own life. Cancer cells show an autonomous drive to proliferate, while ignoring all growth-inhibiting factors. Reminiscent of the immortal cell lines we just covered, cancer cells can "keep dividing endlessly without exhaustion or depletion, generation upon generation." (Mukherjee 2010, p. 458)

Before going into more detail about various forms of cancer, one way to make this topic more tangible for the students is to go through a somewhat farfetched exercise, where we compare the school building we are in to a living organism. The building, we know, was designed by an architect, whose vision was then set down as a blueprint that guided the construction team as it turned the architect's idea into a three-dimensional reality. Once built, the building stands on its own, and is only in need of occasional repairs as it ages. If we imagine, on the other hand, that the building is alive, then things get more complicated. Leaving aside the developmental phases that an actual organism would go through, we still have the challenging situation that the building materials are alive, that they are not only composed of cells that die and need replacement on a regular basis, but that this process needs to take place in sync with the architectural plan of building. (We can imagine, for simplicities' sake, that bricks make up most of the building and that these are constantly dividing, but in such a way that the walls not only remain straight, but also maintain the same height and thickness.)

In other words, the architectural idea must be present and active in all the cells as long as the building exists in order to guide this process "according to plan."

Not only the bricks are cellular, of course. The building's other "cells" have also gone through a differentiation process in one direction or another—into glass cells where the windows should be, into doorframe cells around major openings in the walls, into chalkboard cells in the front of the classroom, etc.

This "living building" will, if it remains healthy, meet us every day with the same hallways, stairways, labs, etc. If we picture the room we are in at the moment, we notice how windows, doors, walls and floors maintain the same dimensions year in and year out. However, if the room were alive, the actual substance they are made of (the cells) would undergo constant renewal—although with individual differences in tempo:The floor cells will need to divide more rapidly, we conclude, than the window cells, and so on.

But then, one day we notice that the door will barely open—the door frame cells have been multiplying too rapidly and are making the opening for the door smaller. So we decide to just leave it open for the time being. But soon we discover that the wall at the front of the classroom is bulging out into the rest of the room, decreasing the room-size, requiring the teacher to move into the space where the student desks are located, and forcing the students to crowd closer and closer together. In a few days, we realize that the front wall will soon reach the door frame and will close off that opening altogether—so we flee our once beloved classroom before it is too late!

We can take this imaginary situation in different directions, of course. We might want to note how gradually the wall-forming cells are growing less differentiated, are becoming more blob-like, more embryonic in nature. And then, oh horror, we notice that some of those cells break off and roll down the hallways, where they attach themselves in various places and start growing (rapidly dividing)—thereby establishing new wall-segments around the building. Soon the former wall cells have become so formless that no one can tell where they actually came from.

Obviously, many variations of this narrative are possible, but essential is the realization that living cells in a healthy organism are always in sync with the blueprint that is the organizational principal of the living entity in question.

When healthy, *the organism always differentiates as a whole, and the particular cause of the differentiation of a part is determined by its position with respect to the whole.* (Maienschien 2007, p. 117).

This is not a one-time, static relationship, but a dynamic, ongoing one. If the "parts" fall out of sync with the whole, then big problems arise.

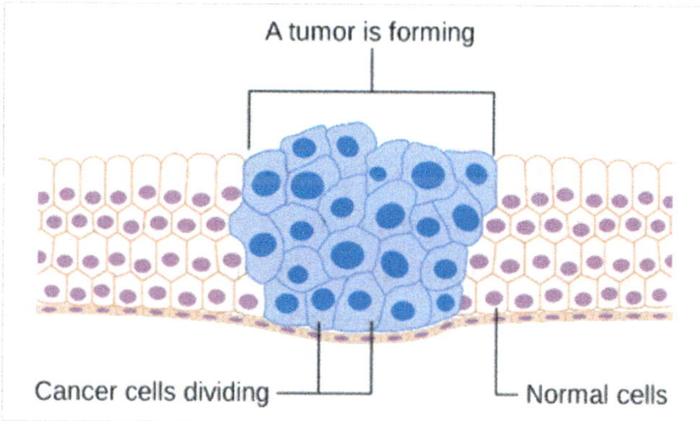

Fig. 1.22 Diagram showing how cancer cells keep on reproducing to form a tumor. (Cancer Research UK. CC BY-SA 4.0 International)

In this fanciful example, they arise when the cells begin to behave as they often do when they are removed from the organism and cultured in vitro: They divide and divide and divide. In the case of cancer, the organism no longer appears able to control and shape "the unfathomable power of cell growth" (Mukherjee).[53] A drawing such as the one above can provide a picture of the kind of changes that take place when cells begin to divide uncontrollably.

The students are familiar with references to "benign" and "malignant" tumors, so it is important to clarify what these terms actually mean. When cells begin proliferating excessively, the abnormal mass that results is called a tumor or neoplasm (new growth). If the tumor remains local, grows slowly and does not interfere with normal body functions, it is termed benign. Malignant tumors, by contrast, grow relentlessly and begin to lose their differentiation as they continue to grow. More aggressive tumors dedifferentiate to such a degree that it is no longer possible to determine the kind of tissue they originated from. The more primitive they become, the more rapidly they divide. These cells are said to be "anaplastic." (Anaplasia means "reversion to a more primitive type" or "to form backward.")

Whereas normal cells die when they lose contact with the surrounding matrix, malignant cells tend to break away from the parent mass and travel vial lymph or blood to other organs, where they develop secondary growths. Such spreading of cancer cells is called "metastasis."

53 Early researchers noticed all the veins and tributaries around masses of malignant tumor cells and thought they looked just like a crab's legs extending outward from every part of its body. From that impression they gave it the name cancer, which is the Greek word for crab.

Since cancers are grouped according to the tissues they are presumed to originate from, the students appreciate hearing briefly about several main categories such as carcinoma, sarcoma, leukemia, lymphoma and melanoma.

After focusing on such details, it is good to wrap up this topic by reflecting again on the contrast between malignant cancer cells and normal cells. The latter develop in harmony with, and in service of, the organism as a whole. By contrast, cancer cells—once "law-abiding citizens"—begin to lose their specificity (develop backward) and not only ignore, but attack the larger whole from which they came. What is normally an exquisitely regulated process of cell division and differentiation, now becomes unbridled growth—an expansionist disease that invades and colonizes other tissues.

In conclusion, it can be interesting to offer for discussion a few farfetched comparisons, such as what it is like if a ship loses its captain, an orchestra its conductor (or its sheet music!), a football team its coach and quarterback, or—heaven help us—a group of students has to spend the day without their teacher! Although such comparisons are inexact, they can spark a lively response from students who play in an orchestra, on a sports team, or are part of any activity that depends on one or two individuals who understand and help organize the whole. After all, the term "organization" implies some kind of agency that coordinates and regulates multiple interwoven components beyond the level of erratic, capricious interaction. (Coleman 2006; Harris 1976; Marieb & Hoehn 2012; Mukherjee 2010)

PART II.

The Mystery of Heredity—The Dramatic Story of a Science Unfolding

Introduction

As the students learned at the outset of this block, human beings have long had questions about the nature of heredity. We took the question of "preformation or epigenesis?" as our starting point for exploring how organisms develop from a fertilized egg to a mature organism. Along the way we found that there is a lawful sequence of developmental stages (morula→blastula→gastrula, etc.). We also saw a certain plasticity in that development (prospective potential > prospective outcome) and how an overriding "organizing principle" is at work in various ways, such as in the regeneration of missing parts and in the healing of wounds. Furthermore, we considered how the uncontrolled development of cancer cells is an indicator that that organizing principle is no longer in command everywhere in the body.

As we move from the perspectives of developmental biology to the study of heredity, it will be important to highlight from time to time how the kinds of questions being asked not only influence the kind of phenomena that researchers are attentive to, but can also lead to a corresponding lack of emphasis on other aspects the organism. This helps the students understand how the development of a scientific discipline can become one-sided over time and may require a course correction (paradigm shift) once its limitations have become too obvious to ignore. The study of genetics and how it unfolded in the course of the 20th century provides an excellent example of this.

Since genetics plays such a central role in our current understanding of the human being, the students are excited to learn more about it. The challenge is how to do so in a way that does not rapidly become abstract and disconnected from their own life experiences, since genetic activity is multifaceted and extremely complex. In fact, by the end of the 20th century it had become less and less clear what, in fact, a gene actually is and does.

> *The more molecular biologists learn about genes, the less sure they seem to become of what a gene really is. Knowledge about the structure and functioning of genes abounds, but also, the gene has become curiously intangible....* (Beurton, et al 2008, p. ix)

In light of such challenges, I have found it most effective to teach modern genetics clothed in the key stages of its historic unfolding. As mentioned earlier, the history of a science is often anything but a straightforward "brick upon brick" story.

20th and early 21st century genetics are a great example of this. The depictions that follow are more extensive than what can be covered in a single block, but the thread that runs through them is an important one. It can easily be sculpted (pared down) to fit the specific class and the time available.

By starting with Mendel and Morgan, and then moving through Watson and Crick to the Human Genome Project and the emergence of Epigenetics, the students learn to see how a specific field of science has progressed. This includes both the defining characteristics at its starting point, as well as the way in which those early perspectives have evolved up to the present day. As will also be evident in what follows, human strengths and human frailties have had a significant influence on the unfolding of this science.

Gregor Johann Mendel[54]

For millennia human beings have noticed that the members of a species are able to breed among themselves and produce similar offspring—from dogs come puppies, from hens come chicks—but the scientific exploration of how that comes about is a much more recent undertaking.

Embedded within the fundamental phenomenon of a unique species-specific organizing principle that is passed down from generation to generation, is the fact that certain "details" (isolated traits) can also vary within the larger whole—such as eye color, hair color, skin pigmentation. It is here, in the realm of the details (trait variations), that the interest of animal breeders was focused. For example, in the 18th century British agriculturists used selective breeding to increase meat yields in cattle. In regard to sheep, they focused on how to select for larger animals that produced longer, more lustrous wool. Despite all the practical knowledge that animal breeders had acquired over the centuries, there was very little understanding of how the mysterious force of hereditary actually functioned.

To rouse the student's interest in this topic, it helps to create the human context in which the discoveries we will be describing came about. In what follows I describe some of what I have found to be of interest to them. For example, they can usually relate to some of the challenges that Johann Mendel—the first individual to formulate and systematically address certain questions regarding heredity—faced, and benefit from hearing how he tenaciously overcame them.

Mendel was born in the small Austrian village of Heinzendorf (now part of the Czech Republic) in 1822. He belonged to a family that—as a last remnant of the feudal system—farmed land that belonged to a noble person, Countess Truchess-Ziel. Although the family owned two horses, money was scarce. As a boy, Johann worked with his father in the fruit orchards, where he learned grafting and helped take care of the beehives. He was also embedded in an environment with many sheep herds, where selective breeding was common.

Because he was a good student and no middle school was available in his village, when Johann completed elementary school he had to transfer to a school in

54 Although anyone who teaches biology will have learned extensively about Mendel and pundit squares, I find the primary importance of covering Mendel is to show the human story behind his research and then to illuminate clearly how the kind of thinking that Mendel used is symptomatic for tendencies that have dominated much of modern genetics since his time.

Fig. 2.1 Gregor Johann Mendel.

a nearby town, which brought with it the additional cost of room and board. Because his family could only afford to pay half his boarding costs, his mother had the local mailman bring food to him when he delivered mail to that town. Despite this, Johann was often hungry. There are many more significant details to his story that I will not expand upon here, but suffice it to say that because of the ongoing financial challenges the family faced in order for him to complete his education, Mendel joined the academically oriented Augustinian Order in 1843. As an Augustinian monk he was thus able to complete his schooling. Joining the order also meant changing his name. He now became Gregor Mendel. After being ordained in 1847, Mendel became a parish priest and worked with the sick and dying. Due to his sensitive and empathetic nature, however, such work made him ill and he could not continue it.

As a result, he applied for a job as a teacher of math and science. Gregor was hired, but with a catch. He was required to take a formal exam that would provide him with a teacher's certificate. He failed the exam. Although he was very learned in the sciences and math, Gregor was a very poor test taker. As a result of the failure, Mendel was required to attend the University of Vienna 90 miles away to pursue a degree in the natural sciences. After two years he returned to the monastery in Brünn and began working as a temporary teacher at the new technical school there. Mendel was popular with the students, who found him to be clear in his presentation of new material, and to be patient and fair in his dealings with them. His deep interest in, and enthusiasm for the sciences was evident. To become a fulltime teacher, however, he still needed to pass the teacher's exam. So, some five years after his first attempt, Mendel returned to Vienna to retake the exam. He failed again! It appears that his sensitive nature once again made it difficult for him to perform well under stressful circumstances.[55]

55 I find it worthwhile to go into this much detail about Mendel's challenges with test taking, because during my teaching career I have witnessed that some of my best students, ones with a great deal of potential in various subject areas, were not necessarily good "test takers." I think it is important for students to realize that the one-dimensional nature of many exams often does not provide an accurate reflection of someone's abilities and their chances for success later. In the case of Mendel, we see an individual—who was to become one of the most famous names in the history of modern science—twice unable to pass a standard teaching exam! Seen from the examination perspective, he was a failure, seen from the perspective of his actual life work, he was a giant! (Which is not to take anything away from individuals who are good test takers.)

Mendel's failure to receive his teaching credentials had an upside, however. It allowed him to focus more of his energy on his own research. As it turns out, three years prior to the second test Mendel had already begun an interesting research project in his tiny monk's quarters. He was breeding white and brown mice to see what color fur their babies would have. Would they be a blended light-brown, all-brown, or all-white (or maybe even polka-dotted)? And how about their kids? Mendel was never able to answer these questions because—so the story goes—the mice he kept in tiny cages in his room began to stink, which soon created a problem. The local bishop happened to visit the monastery one day and was not pleased with the messy, smelly habitat that Mendel's experiments were creating. This was not fitting for a monk and a monastery, the bishop decided, and so they had to go!

Mendel, tenacious as he was, did not give up. He had already overcome many obstacles in his short lifetime (beyond the failed exam) and was not about to let difficult circumstances quash his desire to learn. So he just switched gears. Mendel knew that not only humans and animals passed on traits to their progeny, but plants did too. Thus, he moved his research to the monastery green house (and later to the garden), which not only made his room smell better, but also provided him with a more suitable research object for the kind of experimentation he had been planning. His primary object of study now became peas. (Cobb 2015; Mukherjee 2016; Yannuzzi 2004)

Mendel's Pea Experiments

Mendel gathered 34 pea strains from neighboring farms and crossed them to see which strains bred "true"—for example, short-stemmed plants that generated only short-stemmed plants, tall-stemmed that generated only tall-stemmed, etc. He came up with seven such true-breeding traits:[56]

1. stem length (long vs. short)
2. pea pod texture (full, inflated vs. constricted, wrinkled)
3. pea pod color (green vs. yellow)
4. flower location (terminal vs. axial)
5. flower color (white vs. purple)
6. seed color (yellow vs. green)
7. seed texture (smooth vs. wrinkled)

56 It is important to note here that Mendel included only traits that showed very little variation. Characteristics such as leaf shape or size, for example, varied too much and were excluded from his studies. Another example of this was the color of bean flowers. He found that when he crossed red and white flowering plants, the next generation showed various color gradations between red and white (blending). (Vallejos 1994)

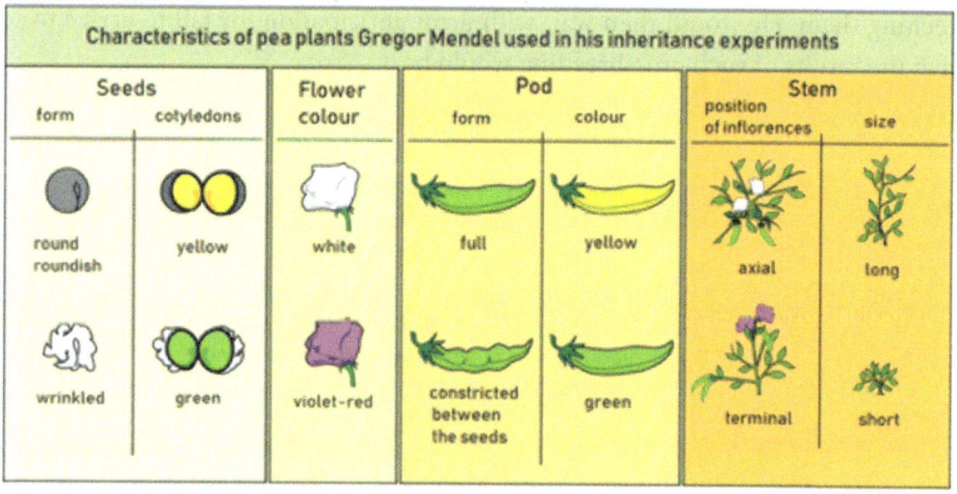

Fig. 2.2 Characteristics of pea plants Gregor Mendel used in his inheritance experiments. (LadyofHats. CC C0 1.0 Universal)

Every trait came in two or more subtypes, but in each case Mendel chose only two (seen in the parentheses above) with which to experiment. The next big step for Mendel was to see what happened when he crossed two purebred subtypes of the parent generation (P), such as pure white-flowered plants with pure purple-flowered ones or pure tall-stemmed plants with pure short-stemmed. To produce this next generation (the F1 generation[57]) of hybrids without inadvertent mix-ups in pollination was, however, no small deed. (Some classes find it interesting to speculate how he would need to go about this before they learn how he actually did it.)

If we take one of the seven trait-subtypes—white vs. purple flowers—as our example, then Mendel's undertaking looked like this: When springtime came and the white and purple flowered peas blossomed, Mendel would usually enter the garden in the early morning—when there was less wind and fewer insects were present—with tweezers in hand. He would move from plant to plant, gently opening the flowers to reveal the stamen, from which he would snip off the pollen bearing anthers and drop the pollen sacs in his jacket pocket—pollen from white flowers in one pocket, pollen from purple flowers in the other. Later he would return to these flowers, and with the help of a small paint brush, he would dust pollen that came from white flowers onto the stigma of violet flowers and vice versa. He would then place a small cotton bag over each flower to prevent pollen from other sources

57 F1 stands for first Filial generation, from the Latin "filius," which means "son."

reaching them. He would then wait with great anticipation for fall to arrive to see what the results of such crossbreeding would be.

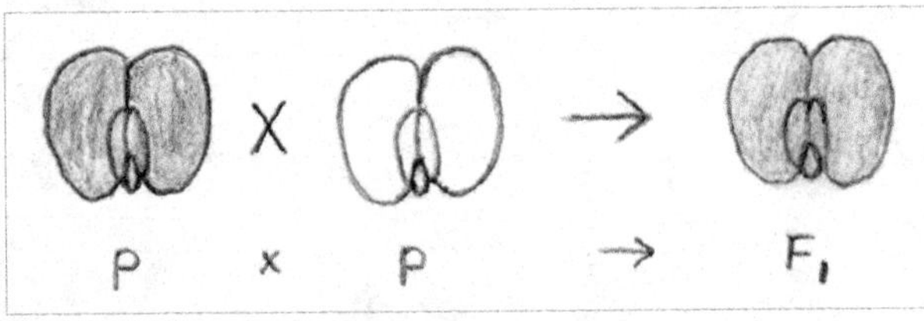

Fig. 2.3 Crossing a purebred purple-flowered member of the parent (P) generation with a purebred white flowered member of the P generation results in a next generation (F1) that is all purple.

At this point we can ask the students what they would expect the result to be. Well, to the surprise of most, the new F1 generation was all purple. The same thing happened with all of the hybrids: All of them showed only one of the two possible traits. All of the F1 generation had tall stems, yellow seeds that were smooth, inflated pods and flowers on the stem. So what happened to the other traits: the white flowers, short stems, wrinkled green seeds, crumpled pods, and flowers that grow at the tip? Half the possible variations disappeared over night (over summer). Now he had a bonafide mystery to work on!

So what next? Do we accept the mystery and move on, or are we like Gregor Mendel and want to dig deeper? If Mendel didn't stop here, we ask the class, then what did he do next? It is not too hard to imagine his next step. Cross those just crossed and see what happens. So, once again, careful surgery, dusting and a long wait. When the grandchildren (the F2 generation) finally began to flower and fruit, Mendel made a surprising discovery. Most of the plants showed the traits that were found in the F1 generation, but a significant minority did not. Although most were purple flowered, tall-stemmed, there were now a fair number of white flowers, short stems, too.

But Mendel doesn't just leave it at that. We have already seen what a careful, organized investigator he is, so it does not surprise the class to learn that he takes a further step—something very unusual for a naturalist in his day: He applies math and does a statistical analysis of his findings. He very carefully counts how many members of the F2 generation show the traits seen in the F1 generation and how many show the traits that were only present in the other half of the P generation. Mission accomplished? For most of us, it would now be time to take a break in the shade and enjoy one of those ripe red apples glistening in the orchard next door.

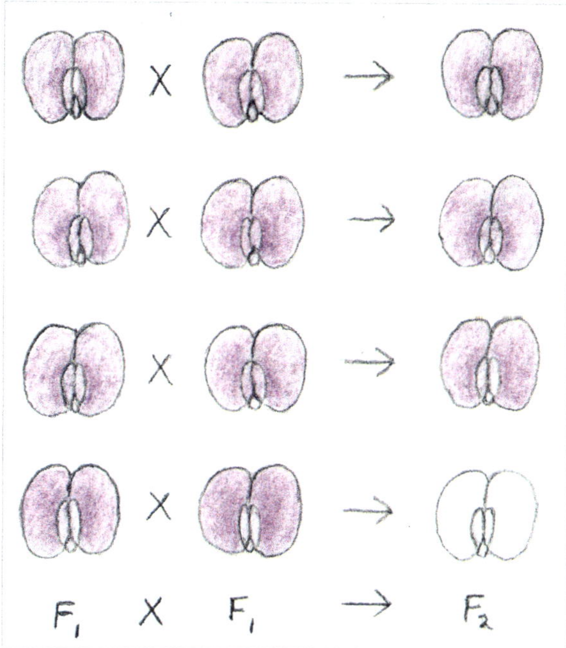

Fig. 2.4 Crossing purple flowered members of the F1 generation results in a 3:1 ratio of purple to white flowers in the F2 generation.

But not if you are Gregor Mendel. Once Mendel had reached the kind of clarity just mentioned, he would pull out more of his purebred seeds and start over again. After all, he asked himself, were his results the real deal, the expression of a natural law, or were they just coincidental. No one else was asking this question of him—no one else was really interested in what he was doing—but he was asking it of himself. He was not willing to stop with an interesting result and then simply generalize it; he wanted to "know," and to know for sure, which meant testing for this outcome again and again and again.

With time Mendel's thumbs grew numb from shelling, and his notebooks overflowed with tables and data from thousands of crosses. Between 1857 and 1864, his small plot of land in the monastery garden produced an overwhelming amount of data to be analyzed: 28,000 plants, 40,000 flowers, and almost 400,000 seeds! (Holdrege 1996; Krogh 2010; Moore 2003; Mukherjee 2016; Yannuzzi 2004)

Mendel's Findings

What did the data show? One central finding was the carefully-counted outcome of crossing members of the F1 generation, as described above. Whereas the F1 had been uniform with one trait dominating, the F2 generation brought forth the following (Krogh 2010):

- 705 purple flowers and 224 white flowers
- 787 long stems and 277 short stems
- 5474 smooth seeds and 1850 wrinkled seeds
- 6022 yellow seeds and 2001 green seeds
- 882 inflated, full pods and 299 wrinkled, constricted pods
- 428 green pods and 152 yellow pods
- 651 flowers on the stem (axial) and 207 at the tip (terminal)

After the teacher puts these findings on the board, the students can be broken up into small groups to see if they can discover any patterns in this data, any lawful relationships. Clearly one trait dominates over the other in these pairings. The mathematically inclined will soon notice—we hope—that the degree of dominance is very similar in each case. Doing the math, they discover that the ratio of the dominant trait to the other (soon to be called "recessive") is as follows and can be added to the data on the board.

- Flower color – 3.14 : 1
- Stem length – 2.84 : 1
- Seed smoothness – 2.96 : 1
- Seed color – 3.01 : 1
- Pod texture – 2.95 : 1
- Pod color – 2.82 : 1
- Flower location – 3.14 : 1

Clearly there is a 3-to-1 relationship between the pairs of traits that Mendel was investigating, which unveils a lawfulness that no one had ever noticed before. In doing this, Mendel confirmed something that one of his favorite professors at the University of Vienna, Christian Doppler (discoverer of the well-known "Doppler Effect") had impressed upon him: that at work within the complex—and sometimes chaotic appearing—phenomena of the natural world were highly organized, but often hidden, natural laws. Doppler's demonstrations and experiments at the university had captivated him, and now, as the fruit of seven years meticulous effort, Mendel found himself in the position to unveil such lawfulness in the realm of heredity.

It is possible to go on at this point and show further examples of the mathematical patterns Mendel found when he bred plants for two traits. (Which is called a dihybrid cross and resulted in a 9:3:3:1 ratio.) However, in my experience a more fruitful alternative is to step back at this point and show how Mendel went on to assess the findings described above. (Carlson 2004; Holdrege 1996; Krogh 2010; Moore 2003; Mukherjee 2016; Yannuzzi 2004)

Mendel's Thought Model

It is important for the students to learn that Mendel did not simply accept the mathematically consistent, and thus lawful nature of his findings. He wanted to understand "why." To this end he created a thought model that would explain the outcome of his experiments. He postulated that for each trait he studied the possessor of that trait had received one "hereditary factor" (later to be called an "allele') from each parent organism. Such factors had never been observed by Mendel nor anyone else, they were a pure thought-construction on his behalf. However, if one assumed that these factors were active "behind the scenes," then Mendel's findings would follow with necessity and the ratios he found would make sense.

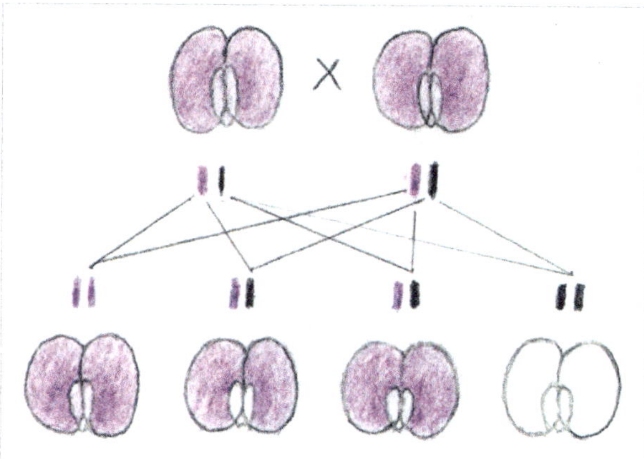

Fig. 2.5 Illustration of Mendel's thought model applied to flower color.

Taking flower color as our example, something like the diagram seen above can be drawn on the board to illustrate Mendel's thinking. Following the thought model, each member of the P generation is shown with two factors that they received during pollination, one from each of their parents. Each member of the F1 generation has one white and one purple factor, which makes them hybrids. That all the flowers turn out to be purple is a strange outcome, unless one makes, with Mendel, a further deduction: namely that when the two factors are present together, one dominates and the other recedes. In this example, the purple color is dominant, so the F1 generation is "uniform" in its purple color.

When we proceed further and cross the members of the purple F1 generation, Mendel's model leaves us with an interesting new situation. The 3:1 outcome purple to white that we can see with our eyes (the phenotype), would appear to be possible because—as we see in the diagram below—the combining of the hereditary factors (the genotype) would lead to three possible combinations: a purebred purple, two hybrid purples, and one purebred white flower (1:2:1). (Krogh 2010; Moore 2003; Mukerjee 2016)

The Reaction to Mendel's Findings

After discussing and clarifying further this process with the students, we can introduce a variety of terms used today that help us describe what is happening. Terms such as alleles, dominant and recessive, homozygous and heterozygous, phenotype and genotype, are helpful in this way.[58] (I save the term "gene" for later, even though it is contained in one of the expressions above and is familiar to all of the students.)

In February 1865 Mendel presented the results of his research to the Natural Science Society in Brünn. About 40 people attended. His paper contained dozens of tables and symbols to denote traits and variations, but for the biologists present it awakened little interest. Why? They normally studied morphology not statistics. As one historian summarized the situation:

> *The counting of variants in seeds and flowers… must have mystified Mendel's contemporaries; the notion of mystical numerical "harmonies" lurking in nature had gone out of fashion with Pythagoras.* (Mukherjee 2016, p. 53)

His 44-page paper was published in the Society's annual proceedings and—following normal procedure—copies sent to dozens of institutions, including the Royal Society in England and the Smithsonian in Washington. Mendel himself received 40 reprints that he sent out to numerous scientists. The response? Between 1866 and 1900 the paper was cited only four times. It had essentially disappeared from scientific literature and his name and work were lost to the world.

Between 1867 and 1871 Mendel continued his experiments with a different kind of plant, the Hawkweed, and was unable to find the kinds of patterns he had observed in peas. Soon thereafter he was promoted to the position of abbot in the

58 All of which were coined post-Mendel.

monastery and had so many administrative responsibilities that he could no longer continue his plant studies. When Mendel passed away in 1884, a young monk described him as "gentle, freehanded, and kindly.... Flowers he loved," and when the local newspaper wrote an obituary for Mendel, not one mention was made of his experimental research! (Krogh 2010; Moore 2003; Mukerjee 2016)

Obvious questions for the class are: a) Why was there so little interest in Mendel's findings, and b) if that is the case, why have we spent so much time learning about him?

Mendel's Approach to Nature Was New—It Awakened Little Interest

It is time to step back and look more consciously at Mendel's research method and mindset. His way of investigating nature was so new that his findings awakened little interest when they were made public. At the same time, paradoxically, his method stands out as a harbinger of a new way of studying nature that was to unfold with vigor in the upcoming 20th century.

What we see in Mendel's way of working is a huge shift away from the "embryological/developmental" approach to nature that occupied us at the outset of this block. We can now ask the class to compare these two approaches. It is helpful to have them break into small groups and—with the assistance of their block books—look for key findings in the developmental biology section of the book that can be compared with Mendel's way of working.[59]

Review

Some findings from our earlier look at the embryological/developmental approach were:

• Developing embryos do not develop the way lawn mowers are built—they are not put together piece by piece. Instead, there is at first ongoing subdivision of the zygote into smaller and smaller components (blastomeres) and then radical changes in those components: They differentiate into a wide variety of cell types that become organized spatially into tissues and organs (morphogenesis).

• When transplanted, developing tissues have the potential to take on the characteristics of their new location ("location correspondent" development)

59 This provides a great opportunity to review earlier parts of the block and widens the frame of reference with which one can view Mendel's work and what will follow there from. (It is not relevant at the moment that Mendel's research took place prior to the developmental findings they are being compared with.)

rather than sticking to what they would have become in their original site ("origin correspondent" development)

• Rather than being a mosaic of predestined parts ("mosaic theory"), we saw how the developing embryo retains significant flexibility ("regulation theory"), with the potential to develop in different ways depending on outer conditions. For example, each half of a two-celled embryo was able create an entire embryo when those cells were separated. We saw how the "Prospective Potential" of each part of the embryo was larger than its "Prospective Outcome."

• Schotte's experiments showed how frog flank tissue could turn into a salamander head (despite retaining certain frog characteristics—no teeth!).

• Hadorn's experiments showed the amazing regenerative capacities of an organism, how—amazingly—cells have the capacity to "trans-differentiate" if need be in order to fit into the overall body plan.

• All in all, development biology points us strongly away from the idea of discrete unchanging traits (units) and toward plasticity and context-shaped characteristics.

* * *

Through this review and further discussion, we can return to Mendel with an expanded awareness of the kind of approach that he took. We can see more clearly what was at the center of his attention, and also what he was overlooking (consciously excluding). For example, at the beginning of his research Mendel obtained 34 varieties of pea plants. From these he selected 22 that revealed constancy in their individual traits, and from there he trimmed the number down to seven traits. They were chosen because they fit his intentions best: They provided discrete traits with no ambiguity. The naturalist recognizes the essence of the plant world in its dynamic unfolding through various stages (from seeds to roots and stem, to leaves, calyx, flowers, fruit, and back again to seeds…). Gardeners and ecologists are interested in how plants react when their access to sunlight, moisture and nutrients changes (phenotypic plasticity). Mendel, by contrast, sought out and focused his full attention on certain discrete characteristics that did not vary, come heck or high water.

Through such reflections, it becomes clear that instead of studying the organism as a whole—in its temporal unfolding and in the dynamic relationship of the parts to each other—Mendel dealt with the plant as a collection of discrete traits and then narrowed his attention even further to just a small number of those traits—ones that stood out clearly and changed very little. Once he restricted his attention to this limited selection of phenomena (purple or white flowers, tall or short stems, etc.), he was able to quantify how many times each variable appeared and when (in which generation). In this reduced field of quantifiable phenomena, clear statistical patterns were discernible.

As discussed earlier, after achieving this kind of mathematical clarity, he could step back and create a "thought model" that explained such statistical patterns. It, too, called for discrete units (factors) that did not interact with each other, and did not go through plant-like transformations. They simply coexisted side-by-side (unmerged), and, at times, separated randomly.

Once we have discussed all this, we can summarize with the class three aspects of Mendel's methodology that were new in their application to living organisms. These were so new, as we saw earlier, that they awakened next-to-no interest and understanding in the community of biologists that were privy to his research. This did not change for the next 34 years following their publication. Briefly summarized, those three aspects were:

1. He focused on isolated traits, not the organism as a whole.

2. He analyzed his experimental findings mathematically.

3. Based on his mathematical findings, he created a thought model that postulated discrete hereditary units (factors) that stand behind the sense-perceptible traits and makes it possible to explain why the mathematical ratios appear as they do. (Holdrege 1996; Hubbard 2013; Krogh 2010; Moore 2003).

Mendel Discovered

What happened to break the 34-year silence around Mendel's work, what made him suddenly relevant? Until the 1870s, there were no methods available to reveal cell contents beyond the largest components such as the nuclei. But with the creation of stain technology this changed. In the late 1870s—over a decade after Mendel's paper was published—Walther Flemming observed live cells dividing in salamander tail fins, and now, with the help of the newly developed dye staining techniques, saw how shadowy masses of threads formed when the nucleus dissolved. These threads that absorbed the dye—and that would soon be called chromosomes (Greek for colored bodies)—captured his attention:

> *One longitudinal half of each thread could move into one half of the nuclear figure, the other longitudinal half of the thread into the other half of the nuclear figure, in other words, each into a future daughter cell.* (Flemming 1880)

What had he discovered? The class should know: mitosis (mitosis = thread)! A few years later, the Belgian cytologist Edouard van Beneden discovered that in the nuclei of the egg and sperm cells (gametes), the number of chromosomes was reduced by half, so that when they merged, the fertilized egg cell (zygote) possessed the normal number of chromosomes, and not double that amount. With these observations, an empirical basis was given for Mendel's thought model.

Although this breakthrough was waiting in the background, it was not until the year 1900 that—seemingly out of the blue—three researchers discovered Mendel's findings, and they did so quite independently of each other. All three were doing research on plant hybridization. In the course of a literature review, they each came across Mendel's paper from 1866 and were astonished: Not only was Mendel's research more exhaustive than their own, but he showed with great mathematical clarity what each of them was just beginning to discover nearly 40 years later. Add to that the fact that Flemming's and van Beneden's empirical findings, which showed a cytological correspondence with the thought model, and things were rapidly beginning to fall into place! The consequence was that in the year 1900, all three scientists—de Vries in Holland, Correns in Germany, and Tschermak in Austria—independently published a paper in which they referred to Mendel and his findings. (Harp 1972; Raven 2016; Cobb 2015)

Sutton and the Lubber Unite Heredity and Cytology

Almost overnight, Mendel's explorations became a subject of active discussion and much disagreement. Then came a decisive publication, one that was to move Mendel's way of thinking to the forefront of what was soon to become a new scientific discipline, which would soon bear the name *Genetics*. In 1902–1903, Walter Sutton—formerly "a grasshopper collecting farm boy from the prairies of Kansas" (Mukherjee), and now a PhD student at Columbia University in New York City—published a paper about his cellular-level microscopic observations of lubber grasshoppers, creatures that had been crawling into his overalls and down his shirtfronts for years on the farm back home.[60] In the course of his observations, Sutton found that there were 22 lubber chromosomes that could be grouped in 11 distinct pairs in a progression from small to large. He was familiar with Mendel's findings and soon saw a connection between them and his lubber observations.

Since the 11th grade is already familiar with the basics of both cytology and Mendel's findings, they can now discuss what they think Sutton saw as the connection between his and Mendel's research. After gathering reflections on this, we provide them with the concluding words of Sutton's paper from the fall of 1902.

> *I may finally call attention to the probability that the association of paternal and maternal chromosomes in pairs and their subsequent separation during the reducing division as indicated above constitute the physical basis of the Mendelian law of heredity.* (Sutton 1902)

One obvious problem remained in connecting trait inheritance to chromosomes, however. The teacher can try to tease this rather obvious discrepancy out of the class, and with a few hints they will easily get it.

Sutton himself addressed this matter four months later in 1903. In the spirit of Mendel, he drew a logical conclusion about something that could not be directly observed. He concluded that if the hereditary factors (allelomorphs, alleles) are inherited by means of the chromosomes, and if there are far more distinct hereditary

60 Although they don't fly, lubbers have been known to jump 9 feet in a single leap! As an undergraduate, Sutton had sent a lubber in the mail to his professor at the University of Kansas as a prank, but as it turned out, the grasshopper's cells were huge and provided a perfect object for cytological investigation. The famous cell biologist Edmund Beecher Wilson later described the lubber as "one of the finest objects thus far discovered for the investigation of the minutest details of cell structure."

traits than there are chromosomes, then it follows logically that there must be multiple hereditary factors connected to each (or some) of the chromosomes.

> *We have seen reason, in the foregoing considerations, to believe that there is a definite relation between chromosomes and allelomorphs*[61] *or unit characters but we have not before inquired whether an entire chromosome or only a part of one is to be regarded as the basis of a single allelomorph. The answer must unquestionably be in favor of the latter possibility, for otherwise the number of distinct characters possessed by an individual could not exceed the number of chromosomes in the germ products; which is undoubtedly contrary to fact. We must, therefore, assume that some chromosomes at least are related to a number of different allelomorphs. If then, the chromosomes permanently retain their individuality, it follows that all the allelomorphs represented by any one chromosome must be inherited together.* (Sutton 1903)

"Beads on a string" soon became a common image to describe this. Just as a necklace carries multiple beads on one string, each chromosome was now pictured as being the bodily basis for multiple hereditary factors. This view soon became known as "the chromosome theory of heredity." As Ruth Hubbard (2013, p.18) noted: "At a time when invisible atoms, electrons, and quanta were being accepted into the world of chemistry and physics, biologists had little problem accepting that heredity also was mediated by invisible material particles." In the following years, numerous expressions were used for Mendel's invisible particles (factors), until in 1909 the Danish botanist Wilhelm Johannsen coined the term "gene,"[62] which thereafter became the accepted name for these mysterious factors.[63] (Cobb 2015; Holdrege 1996; Moore 2003; Mukherjee 2010; Schwartz 2008; Tagarelli et al. 2003)

61 aka alleles (MH)

62 From the Greek *genno*, "to give birth."

63 Johannsen did not mean to imply actual physical substantiality with the term gene, but some kind of organized predisposition (*Anlage* in German).

The Birth of a New Scientific Discipline: Genetics

With the extension of Mendel's way of thinking into the cytology-based chromosome theory of heredity—as well as through concepts such as Johannsen's "genotype" (subsensory hereditary factors) and "phenotype" (visible traits)—the science of "Genetics" was born. The contrast between Mendel's approach to nature and the perspectives of developmental biologists grew even greater when the hypothesis was advanced that Mendel's factors had an actual physical presence in the form of chromosomes. In other words, the research focus and questions of the geneticist were now far removed from the those of conventional biologists. One prominent historian of molecular biology, Michel Morange (1998, pp. 17–19), characterized the situation as follows:

> *Genetics rapidly developed into a separate discipline that had little contact with other branches of biology. Far from being a handicap, the institutionalized isolation helped the new science, particularly in the United States.... The success of genetics shows that the separation of these topics—the separation of heredity and embryology—was necessary; at least for a time. Because of its intellectual isolation, many biologists considered genetics to be an abstract science that was somewhat removed from reality. Geneticists studied genes without apparently being interested in either the way they worked or their chemical nature.... Geneticists did, however, believe that genes had a fundamental and unique property: self replication. This capacity was inherently linked to the nature of genes, but could be studied independently of their functioning."*

Years later, one key player in the development of modern genetics, Alfred Hershey, put it this way: "As long as you're thinking about inheritance, who gives a damn what the substance is—it's irrelevant." (Cobb 2015, p. 65)[64]

T.H. Morgan, His Team, and "the Fly Room"

The next big step in this new approach to understanding heredity also took take place at Columbia University, this time under the leadership of T.H. Morgan. Before describing in more detail what happened and what was discovered there, one way to quickly awaken the student's curiosity about this foundational step in

64 Ironically, Hershey would eventually become a central figure in shedding light on the substance question.

the development of genetic research is to read a short description of life in "the fly room" at Columbia (Mukherjee 2016, p. 97):

> *Between 1905 and 1925 the Fly Room at Columbia was the epicenter of genetics, a catalytic chamber for the new science. Ideas ricocheted off ideas, like atoms splitting atoms. The chain reaction of discoveries—linkage, crossing over, the linearity of genetic maps, the distance between genes—burst forth with such ferocity that it seemed, at times, that genetics was not born but zippered into existence. Over the next decades, a spray of Nobel Prizes would be showered on the occupants of the room: Morgan, his students, his student's students, and even their students would all win the prize for their discoveries.*

Fig. 2.6 The Fly Room. **A.** Thomas Hunt Morgan, **B.** Alfred Sturtevant, **C.** Hermann Joseph Muller, and **D.** Calvin Blackman Bridges. (Caltech Archives)

T.H. Morgan seemed at first an unlikely candidate to lead this explosion in genetic research. Previously, as a highly respected professor of Zoology at Columbia, he had been skeptical of Mendel's theory of heredity. It took the findings of Sutton and others to convince him that the pursuit of such matters was worthwhile. When he did, finally, make the decision to explore the—soon to be called—"genetics" question, he was immediately faced with the pivotal decision: What kind of organism should he use for his research? On the one hand, he could use plants as Mendel had effectively done, but he would then be tethered to the seasons and could only gather, at best, data on two generations per year. The animals typically used for such studies—mice, rats, chickens, rabbits, guinea pigs, hamsters—were messy to work with due to cage cleaning, etc. He needed organisms that would be inexpensive to maintain, breed rapidly, and produce lots of progeny.

The students usually enjoy reflecting on what they think would be a good choice for such an undertaking. It makes sense to them when they learn that Morgan chose a creature they are all familiar with, one that they may have trouble keeping control of in their kitchens at home during the warmer months of the year. He chose the common fruit fly (*Drosophila melanogaster*). After letting the students suggest some of the reasons why Morgan found it suitable for the task, we list a few.

- Females lay up to one hundred eggs per day
- Eggs develop into adults in ~10 days
- Females are sexually mature 8–10 hours after hatching
- At 3mm they are small enough to easily store in milk bottles
- At 3mm they are large enough for traits to be seen without a microscope
- They possess a large number of distinguishable traits
- They are easy and cheap to feed
- They possess only four chromosomes
- They are "happy" at room temperature

Morgan started breeding Drosophila around 1905. His first stock—so the story goes—came from a flock of flies above a pile of overripe fruit in a grocery store.[65] A year later he was breeding maggots by the thousands. They were kept in milk bottles that he brought from home and filled with rotting fruit. With bunches of overripe bananas hanging from sticks, the smell of fermenting fruit in the lab was overpowering. Ever time Morgan moved, a haze of flies that had escaped would lift off the tables like a buzzing veil.[66]

65 Another story his students often told was that the first flies came from a pineapple sitting on a window sill outside his lab.

66 If you time this class to be during fruit fly season, it is fun to breed a bunch during the course—and the students can all take some home when the block is over! In some classes, it is even fun to have a fruit fly-breeding contest among the students. They breed them at home and on a specified date everyone brings in their fruit fly container for counting. The student with the most fruit flies then receives a free one-week supply (7) of overripe bananas.

His third-floor lab at Columbia was about the size and shape of Mendel's garden and was called "the Fly Room" by students (and historians). As Mendel had done, Morgan began identifying heritable traits that he could track over generations and to catalog those traits that varied, such as eye color, wing shape, body color, body size, bristles, leg shape, etc. He was joined by a flock of students—some of whom would make significant discoveries of their own—who helped him monitor the changes from generation to generation. Already in the first three years they crossed thousands of fruit flies to create tens of thousands of progeny. (Carlson 2004; Krogh 2010; Linder 1998; Moore 2003; Mukherjee 2016; Schwarz 2008)

Linkage Groups Discovered

When the crosses were tabulated, Morgan discovered something very surprising. In contrast to Mendel's findings, where each trait was inherited independently—thus making all combinations of traits possible—Drosophila's traits appeared to be linked together in groups. For example, sable color was always inherited together with miniature wings. That traits were linked in this way related to something that Sutton had postulated in 1903: If there are many more traits than chromosomes, and if chromosomes are the carriers of hereditary factors (i.e., genes), then there must be multiple factors per chromosome. Morgan's findings appeared

Fig. 2.7 Drosophila wing variations.
(Biodiversity Heritage Library. Public domain)

to confirm Sutton's conclusion. Like beads on a string, certain traits appeared to be linked to each other on the same chromosome. These were called "linkage groups."

If this explanation was to be considered valid, however, one other condition had to be met. We open this up to the class to see if they can determine what that might be. Before long, perhaps with hints from the teacher, it becomes clear: the number of linkage groups would need to match the number of chromosomes. And indeed, they did! Morgan's team found four linkage groups and Drosophila has four chromosomes!

But Linkage Groups Regroup—How Can This Be Explained?

However, once that became clear, they noticed something previously unobserved: From time-to-time, traits that were normally linked together would suddenly unlink and pair with different traits. In other words, the traits that made up the linkage groups would—on occasion—re-constellate. The question for the class now becomes: Does this pull the rug out from under the dominant theory in genetics at that time, or can it somehow be explained within the context of that theory?

Just as we have seen before, much of the "Chromosome Theory of Inheritance" involves linking hypothetical, logical thinking to multiple traits at the sensory level, and various chromosome particulars at the microscopic level. It is a continuation of Mendel's thought-model approach: making sense of copious data with logical deductions that draw heavily on mathematical analysis.

Once again, we can turn to the class and ask them to put themselves in Morgan's shoes; to try as a class to do what Morgan and his team did: a) think out possible explanatory scenarios, and then b) see if any phenomena can be found that fit the new thought model. For step a) the students can apply their own thinking. For step b) they must rely on what the cytologists were able to observe at that time.

Thinking this through, it is clear: If the linkage groups do not remain constant, then there must be some kind of process whereby genes (factors) that were on one chromosome are able to migrate to different chromosomes and vice versa. Morgan called such a process "crossing over." The big question then became: Is there any evidence from the cytological side that would support this view? Morgan did a thorough search of the cytological literature and found in the work of the Belgian F.A. Janssens a phenomenon that might confirm his "crossing over" idea. Janssens had observed under a microscope that during the early stages of meiosis chromatids

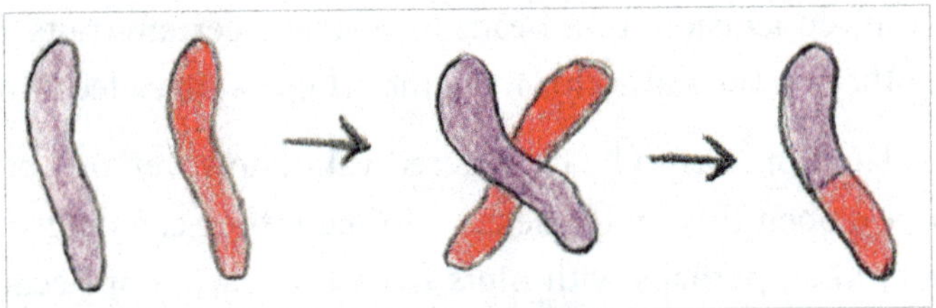

Fig. 2.8 Illustration of Morgan's postulate that this crossing-over leads to breakage and recombination.

sometimes entwined with each other. This was just what Morgan was looking for. He postulated that the overlapping chromatids swapped sections of themselves by breaking at the location where they crossed over, and by recombining with the corresponding section of the other chromatid, as the drawing above illustrates.

The actual breaking and recombining could not be observed, so this explanation remained hypothetical. (Carlson 2004; Krogh 2010; Linder 1998; Moore 2003; Mukherjee 2016; Schwarz 2008)

Recombination Data Lead to Gene Maps—Clarity at Last!

But Morgan's team did not stop there. Assuming the validity of the crossing over idea, and assuming that genes are arranged in a linear order on the chromosomes (beads on a string), then the possibility should exist to determine the locations of the various genes on the chromosome. But how? There are numerous ways to illustrate this to the class. One possibility is to set up a chain of students holding hands in front of the classroom, and then to call on various students, one at a time, to come forward and break through the chain at some point. The question then becomes: Would it be possible to know where the students are located on the chain just by tabulating how often they got separated?

It is quickly clear that the two students at the ends always get separated if someone passes through the chain. This insight can be further refined. The next-to-the-last the student on the left end will get separated less often from the student end than from any other student in the line, and so on. For fun, we could even give the class a pop quiz at this moment by providing them with data that says how often the students got separated from each other and let them figure out where everyone must have stood.

Returning to the fly room, we learn that by tracking the frequency with which genes separate when linkage groups change, Morgan's team was able to create chromosome maps that indicated the relative location of each gene on a chromosome. The diagram below illustrates this. Antennae length and wing shape, for example, must be at the ends since they always separate. Body color and eye color must be next to each other, since eye color separates from body color less frequently than from any other gene, and so on.

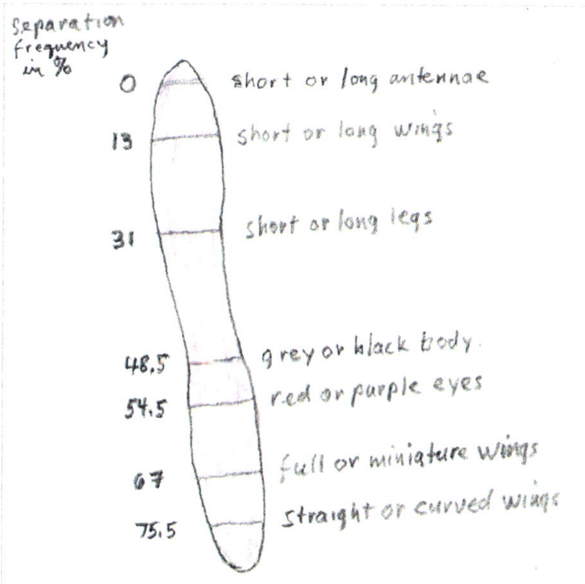

Fig. 2.9 Partial map of Drosophila's Chromosome II. showing the relative separation of genes (alleles) calibrated by the frequency with which they separated from antennae length during meiosis.

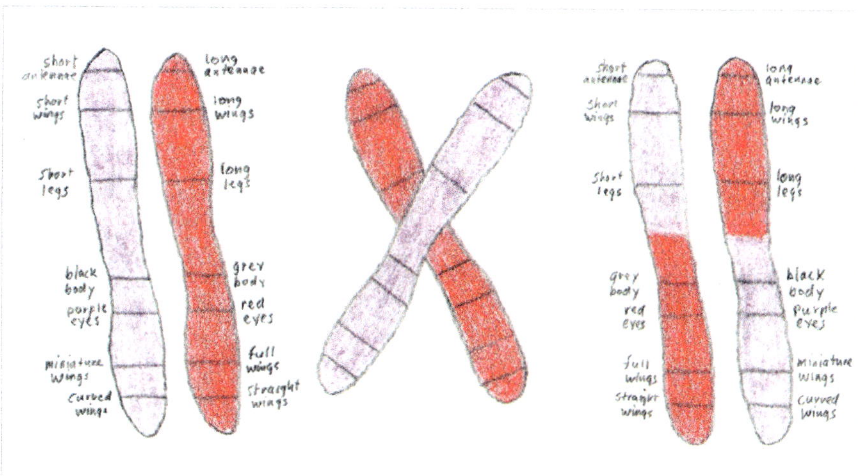

Fig. 2.10 Illustration of how linkage groups can be split by crossing over and new linkage groups formed (genetic recombination).

In sum, tens of thousands of careful experiments that were meticulously recorded and statistically analyzed allowed Morgan and team to create actual maps of Drosophila chromosomes that indicated where the genes they carried must be located. By the time Morgan received the Nobel Prize for this work in 1934, fruit fly research was being carried out all over the world. For that reason, he is considered the founder of what is called "classical genetic research."

Even after all of these discoveries, however, the question still remains: How much was really understood about the actual nature of genes? It is important for the students to see through such examples how the path of science is never simply additive, like placing one brick of new insight onto the next—it is not a mere accumulation of facts upon which ironclad theories can be erected. Instead, there is an ongoing change in the questions and answers pursued by scientists along the long and winding road of scientific discovery. Here, at this juncture, however, it would appear as if we have a firm basis for understanding the transmission of hereditary traits. Chromosome maps had been created showing where the genes for various traits must be located. The creation of such maps was an amazing feat and made things look pretty "ironclad" after all.

But, as so often happens in the history of science, things did not turn out to be quite so simple. Further exploration began to reveal that the "one gene—one trait" link that Mendel spoke of, and Morgan's maps seemed to show, was an oversimplification. A one-to-one linkage between gene and trait turned out to be the exception rather than the rule in most organisms. In what came to be called "polygenetic [multiple genes] inheritance," it was found, for example, that in Drosophila there are 50-some genes that influence the shape and color of the eyes. In the wider world we know, for example, that many genes are involved in the length of an ear of corn, the color of a wheat grain, in the amount of milk that a cow produces. Human traits like height, weight, and skin color also involve multiple genes interacting with each other. In fact, all human attributes involve multiple genes, with the environment playing a role, as well. This can be illustrated for the students with a diagram such as follows.

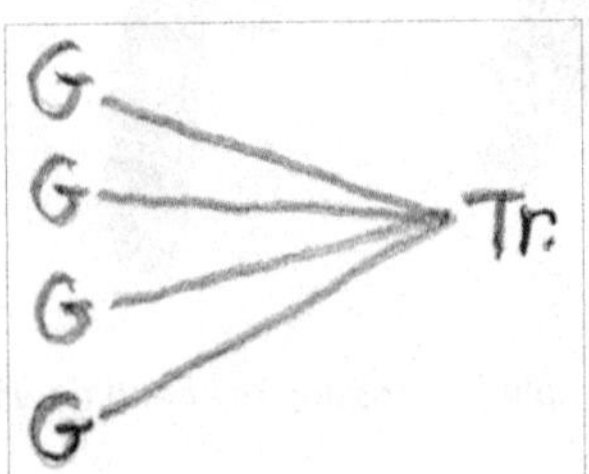

Fig. 2.11 Diagram illustrating how multiple genes (G) can influence one trait (Tr.).

Hermann Muller, who was one of the Morgan's key assistants from the early years in the fly room, and who would himself later receive a Nobel Prize for his work on the physiological and genetic effects of radiation, put it this way:

> *It is a poor elementary course in genetics which does not bring out the fact that each bodily process and part is the resultant of the activity of multiple genes.... For, in the first place, nearly all observed characteristics depend upon many genes. Most of these genes act more or less cumulatively, as "modifiers," so that a change in any one of them usually affects the result only slightly.* (Muller 1950, p.112)

On the other hand, it is also common for a single gene to have many effects.[67] In Drosophila, for example, it was later found that the gene named "Notch" is involved in the formation of the wing edge, the wing blood vessels, eye size, hair growth on the trunk and bristle development on the fly's neck. Know as "pleiotropy" (from Greek: "pleion" = more, and "tropos" = way), the multiple effects of one gene are characteristic of many inherited disorders. (Carlson 2004; Krogh 2010; Linder 1998; Moore 2003; Morange 2001; Mukherjee 2016; Raven 2016; Schwarz 2008)

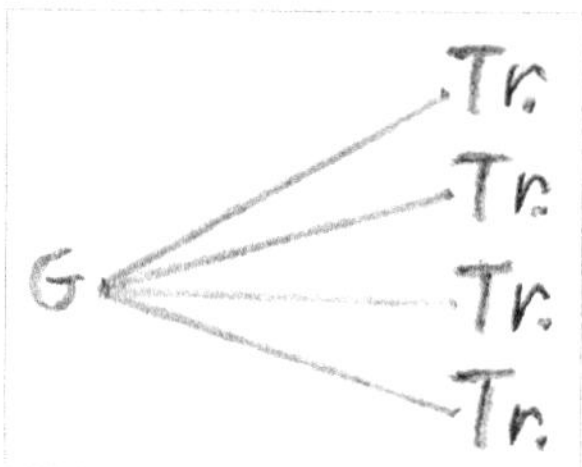

Fig. 2.12 Diagram illustrating how one gene (G) can influence multiple traits (Tr.).

If you put the two diagrams (Figs. 2.11 and 2.12) together, you see a web of relationships—a world far removed from a linear one gene→ one trait causality.

67 A classic example of this is the Notch1 gene. The Notch1 gene is involved in the development of Notch proteins. Although the following is more detailed than would be suitable for most classes at this point in the block, it reveals very clearly the process-oriented, interactive nature of organisms. Notch proteins (together with other protein components) are central in the creation of signaling pathways that make it possible for neighboring cells to communicate with each other. In this way, groups of cells can organize themselves in what are often called "pattern formations." For example, if one cell expresses a given trait, it can be switched off in neighboring cells. In such ways, groups of cells can influence each other during cell differentiation processes and organogenesis. This takes place during both the embryonic and adult phases of life and influences neuron development, angiogenesis, heart valve formation, bone development, and much more. Moreover, Notch signaling has been found to be dysregulated in a number of different human diseases ranging from cancer to neurodegeneration. What this example reveals is anything but a straightforward linear chain of cause and effect. It shows, instead, interactive networking between different parts of the larger whole, thereby enabling them to integrate and subordinate themselves within more comprehensive pattern-forming processes. (Kovall & Blacklow 2010; Lai 2004; MedlinePlus Genetics 2020; Morange 2001)

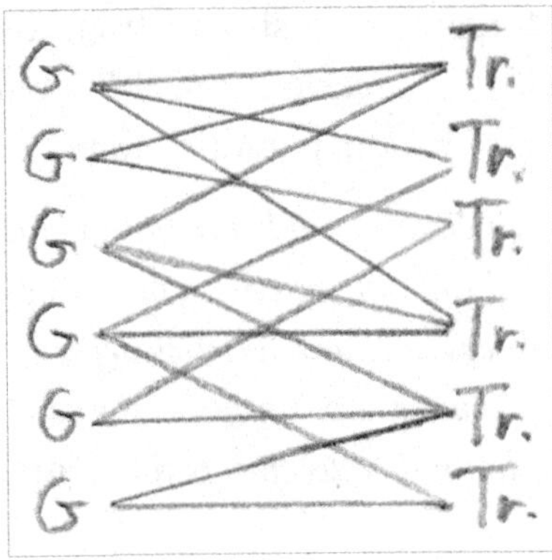

Fig. 2.13 Diagram that combines Figures 2.11 and 2.12 to illustrate a web of causal relationships, which stands in contrast to one-gene-one-trait causality.

Simple Linear Causality Does Not Suffice

But is it really so surprising—we ask the students—that genes normally work together in an interrelated web rather than in a linear "one gene—one trait" fashion? In fact, our biology curriculum in the 9th and 10th grades mirrors in many ways the step from simple linear causality to multileveled, complex interwoven causality. The current block also began with the insight that a clear cut mosaic theory of development is inadequate when trying to understand the flexible dynamic that embryonic development reveals. Instead, concepts like "prospective potential" and "prospective outcome" were necessary to shed light on what was taking place.

After all that we have heard about the genetic research of Morgan and team, the question must be asked: So what, then, are genes actually? Mukherjee (2016) characterizes the situation in this way:

> *It is a testament to the ability of scientists to accept abstractions as truths that 50 years after the publication of Mendel's paper—from 1865 to 1915—biologists knew genes only through the properties they produced.... Dimly, as if through a veil, geneticists were beginning to visualize patterns and themes: threads, strings, maps, crossings, broken and unbroken lines, chromosomes that carried information in a coded and compressed form. But no one had seen a gene in action or knew its material essence. The central quest of the study of heredity seemed like an object perceived only through its shadows, tantalizingly invisible to science.* (pp. 97–98)

In 1933, not long before he received the Nobel Prize for the discoveries made in the fly room, Morgan himself observed: "There is no consensus opinion amongst geneticists as to what genes are—whether they are real or purely fictitious (Morgan 1933, p. 315)." This may seem a surprising statement, but even after all his work on linkage and gene maps, Morgan still had difficulty describing or even imagining genes in a material form.

As we heard at the beginning of this section, the early success of genetics was aided by its separation from embryology and traditional life sciences. Because of its isolation from other areas of biology, genetics was considered by many to be an abstract science that was somewhat removed from reality. Geneticists studied genes, but apparently had little interest in either the way they functioned or their chemical nature.

That worked for a time, but it had now become clear: It would not suffice. Something new was needed.

Biochemistry Joins the Search

Beadle and Tatum

In the late 1930s, one of the scientists who had worked in Morgan's lab, Georg Beadle, resolved to find out something about the actual substance of genes. Now at Stanford University, he worked on changes in the eye color of Drosophila and attempted unsuccessfully to extract the chemicals from the fly's eyes. Such difficulties led "Beets"—as his fellow scientists liked to call him—and his colleague Eduard Tatum to seek an even simpler organism for their research. They settled on *Neurospora*, a bread mold. *Neurospora* is a hardy fungus that is able to survive in the near absence of certain vitamins and amino acids because it is able to synthesize those that it needs—something that most organisms (including humans) cannot do. This is possible because *Neurospora* possesses biochemical pathways that turn sugar, salts, and biotin into all the other building blocks it needs—in particular amino acids and vitamins.

Beadle and Tatum irradiated Neurospora spores with X-rays, thereby creating mutant fungi that were unable to synthesize particular amino acids or vitamins. Since it was known that synthesizing amino acids and vitamins out of precursor chemicals required the presence of enzymes in the cells, they surmised that when Neurospora was unable to synthesize this or that vitamin or amino acid a certain enzyme was now missing. With enormous patience they performed countless laborious experiments of this kind,[68] from which they concluded that each X-ray disrupts the function of an enzyme that is specific to a particular metabolic function. This, in turn, led them to the conclusion that genes must either produce specific enzymes, or be the enzymes themselves. This later became known as the "one gene, one enzyme hypothesis."[69]

68 The students enjoy the following anecdote that illustrates Beadle's prodigious patience and attention to detail: it is said that he once spent an entire afternoon teaching a graduate student how to best marinate a steak: which involved adding one spice at a time, at precisely timed intervals.

69 When Beadle and Tatum published their findings in 1941, they did not emphasize the "one" aspect in the name "one gene, one enzyme hypothesis." In fact, that snappy moniker was first given to their discovery a few years later by a colleague, and did not come from Beadle and Tatum themselves. In their words: *From the standpoint of physiological genetics the development and functioning of an organism consist essentially of an integrated system of chemical reactions controlled in some manner by genes. It is entirely tenable to suppose that these genes which are themselves a part of the system, control or regulate specific reactions in the system either by acting directly as enzymes or by determining the specificities of enzymes. Since the components of such a system are likely to be interrelated in complex ways, and since the synthesis of the parts of individual genes are presumably dependent on the functioning of other genes, it would appear that there must exist orders of directness of gene control ranging from simple one-to-one relations to relations of great complexity.* (p. 500, communicated October 8, 1941)

Had they found the substance that genes are made of? Let's hear how their peers reacted. When "Beets" presented their findings in a seminar at the California Institute of Technology (Caltech) his listeners were stunned. After speaking for only 30 minutes, he stopped. One member of audience later remembered (Cobb 2016, p. 10):

> *We had never heard such experimental results before. It was the fulfillment of a dream, the demonstration that genes had an ascertainable role in biochemistry. We were all waiting—or perhaps hoping—for him to continue. When it became clear that he actually was finished, the applause was deafening.*

The results of their research earned Beadle and Tatum a Nobel Prize in 1958. They had found something that clearly resonated with many biologists and biochemists. One reason for this was the key role that enzymes were already known to play in the life processes of organisms. For that reason, it is important to take a short detour with the students into the world of enzymes. (Beadle & Tatum 1941; Cobb 2015; Morange 2020; Mukherjee 2010)

The Great Accelerators: Catalytic Enzymes

Enzymes were an unknown entity far into the history of the life sciences. It was not until 1897 when, in preparation for an experiment, the German chemist Eduard Buchner ground up some yeast cells (killing them and causing them to break open), before running them through a filter to extract the juice they contained. In order to preserve the liquid from decomposition, he added a thick sugar syrup to the container the juice was in. To his immense surprise, instead of acting as a preservative, the syrup began to ferment with great vigor. The yeast juice, it turned out, was able to convert the sugar into alcohol and carbon dioxide in the absence of actual living yeast cells. Up until then, it had always been thought that alcoholic fermentation only took place in the presence of living cells.

This new mystery substance was given the name "enzyme," which means "in yeast." To their amazement, Buchner and others soon realized that even though enzymes are not living substances themselves, living organisms can only engender fermentative activity (processes in which substances are broken down into simpler substances) when enzymes are present. Further study revealed that enzymes are a certain type of protein.[70] Although enzymes are only a small part of the body's total

70 Although more recently it has been found that under certain circumstances RNA can also function as an enzyme.

protein, they are absolutely crucial and exist in great numbers in every cell. They function as catalysts, which means they accelerate chemical reactions within cells. In some instances, enzymes can accelerate a reaction more than 10 billion times! In humans there are more than 10,000 enzymes (Morange 2020),[71] each of which is specific to a particular chemical reaction.

An everyday example of enzyme deficiency that students are familiar with is lactose intolerance. Millions of Americans suffer from this condition when, after the childhood phase that involves a great deal of milk consumption, their bodies decrease their production of lactase,[72] the enzyme that activates the splitting of lactose into glucose and galactose. As a result, the undigested lactose sits in the intestines and encourages bacterial fermentation, which results in cramps, gas production, etc.

Although they bring about a change in the substance (substrate) that they are specific to, enzymes still maintain their own chemical composition and are ready to pick up another substrate immediately and activate a new reaction. Believe it or not, we tell the students, a given enzyme can activate hundreds, even thousands of chemical reactions in one second![73]

The students are amazed to learn that every moment billions of chemical reactions take place inside each of our bodies. They happen at lightning speed inside our cells. However, if the chemical reactants are removed from the body and placed in testtubes—in vitro— the reactions happen at a snail's pace. Why the difference? Because our cells have the necessary catalytic enzymes within them, testtubes do not.

Enzymes facilitate almost every chemical process that takes place in an organism. No living thing could exist without them. Many metabolic functions require a whole complex of enzymes to complete hundreds of reactions. Although they can be studied individually and often are, they derive their significance from being parts of a larger whole, which is sometimes called the "metabolic web."

---

71 The estimated number of enzymes varies greatly from source to source.

72 It is helpful for the students to learn that the names of enzymes usually have something to do with their functions and end with the suffix "ase." Thus, the name "lactase" for the enzyme that breaks down "lactose."

73 Enzymes sometimes need partners in this process. If the partners are organic—often derivatives of vitamins—they are called coenzymes. Inorganic partners—such as magnesium, iron, potassium and zinc—are called cofactors.

In the words of biochemist Franklin Harold (2001):

> *The metabolic web is a complex system, whose properties as a whole are not always obvious from the properties of its parts and the rules that govern their interactions.*

The metabolic web goes far beyond a simple one-to-one linear causality that was implied in the name "one gene, one enzyme." However, because enzymes are proteins,[74] one of the central impacts of Beadle and Tatum's hypothesis was to give weight to the idea that proteins must play a central role in heredity processes. (Carlson 2004; Harold 2001; Krogh 2010; Linder 1998; Locke 1969; Moore 2003; Morange 2020; Mukherjee 2016; Schwarz 2008)

Ubiquitous Proteins Must Be the Key

Except for water, proteins are the most abundant substance found in human beings. Although the students have heard about proteins in varied contexts since they were children, they are nonetheless surprised when they learn that—if we exclude water—half the weight of the human body is made of protein. Every living creature, regardless of its size, contains a significant percentage of protein.

The name protein means "primary" and is thus a fitting moniker for this substance. In a quick overview, we can explain how proteins perform several major functions in the human body. Structural proteins are part of the bones and connective tissues that provide the body's framework. Proteins that make up the skin, hair and nails provide an outer boundary for the body. Other proteins are found in the muscles and enable us to move. Our blood is teeming with proteins of different kinds, Hemoglobin carries oxygen throughout the body; thrombin, fibrin, etc., are involve in blood clotting, and so on. In short, nearly every cellular function—from metabolism, respiration, cell division, and self-defense to waste disposal, signaling, secretion, growth, and cell death—requires proteins. Proteins are the workhorses of the biochemical world.

In the 1930s chemical analysis determined that chromosomes were composed of protein and nucleic acid (DNA). Such a finding did not imply, however, that the two components were equally important. So when enzymes were added to the repertoire of protein functions—in particular after the findings of Beadle and Tatum—it is not hard to imagine why proteins became the focus of genetic research.

74 It was only determined much later that RNA can play an enzymatic role, as well.

A second reason for this was the chemical nature of the two substances. DNA was understood to be a rather uncomplicated molecule composed of a few bases, phosphate, and a simple sugar. To researchers, such a simple substance did not seem capable of the complex tasks required of genetic material. Many saw DNA as just a glorified scaffolding for genes.[75]

As a result, proteins seemed far more likely to be gene carriers than DNA. Proteins have as their building blocks amino acids. There are 20 different amino acids commonly found in proteins, similar to the number of letters in our alphabet. Like our letters, these can be combined in the most varied ways, although the strings of amino acids are much longer than our words.[76] There are often 300 or more amino acids per protein molecule, with the range typically falling between 50 and 2000. This makes it possible to create an unbelievable number of proteins—more than 100,000 different proteins are now known. What's more, each protein folds into a unique three-dimensional shape. These stable folding patterns make up the secondary structure of a protein.[77]

With one of the two components of chromosomes showing so much more diversity and potential for variation, it was a shut and closed case for most scientists that protein had to be the key player in chromosomes, even if no one was even close to understanding how. The findings of Beadle and Tatum gave this view even more impetus. Nonetheless, as firmly in the saddle as proteins appeared to be, here and there evidence began to appear that pointed in a different direction. (Berg et al. 2002; Locke 1969; Moore 2003; Morange 2020; Mukherjee 2016)

Oswald Avery's Unanticipated Extract

The first serious crack in the protein-as-gene veneer came in the early 1940s from experiments done by the American bacteriologist Oswald Avery and his coworkers at the Rockefeller Institute.[78] They were working with two types of bacteria. One type, the S-strain (smooth coated), was virulent, the other, the

75 This view was given credence by Phoebus Levene, the leading expert on nucleic acids, who referred to DNA as an extremely "unsophisticated" structure; at times he was said to have called it a "stupid molecule."

76 A single 25-letter word composed of 5 different letters can produce 372,529,029,846,191,405 different possible combinations.

77 After such a description, it is important to point out to the students that even though proteins are complex enough and large enough to be called macromolecules, they still exist in a completely different dimension of reality than the one we live in. A single protein molecule is typically a millionth of an inch across—too small to be seen even with the help of a microscope. And we should also remind them that they are not static pieces of sculpture in a mini-world either, but actually part of a metabolic web of constant activity.

78 Avery's research was based on experiments done on S-strain and R-strain bacteria by Frederick Griffith in the 1920s.

R-strain (rough coated), was not. After removing an extract from the virulent S-strain and mixing it with the live bacteria of the R-strain, they injected this mix into mice. The mice died and what emerged were virulent S-strain bacteria. The extract from the S-strain had transformed the R-strain into virulent smooth-coated bacteria that retained these characteristics generation after generation. Which is to say, the S-strain virulence and smooth coat had been transmitted to the R-strain by the extract taken from the S-strain. So, the big question became: What kind of substance is this transformative extract whose effects are inherited from generation to generation? After many long and painstaking experiments, Avery and colleagues came to a surprising, but unambiguous conclusion: The transformative substance was DNA—a nucleic acid!

Avery's findings had an interesting effect. For some they became a hot topic that led to a wave of research into the structure and function of DNA. The well-known chemist Erwin Chargraff, for example, began sketching out ways in which DNA might not be so boring, after all. He hypothesized that if the sequence of bases that it contained were to vary, then the kind of variability that would be necessary for genes might be possible.

Avery's findings were so out of sync with the already-drawn conclusion that protein had to be the key genetic player in chromosomes that many researchers just assumed there had to be errors involved in his results—otherwise it made no sense! One well known scientist, Max Delbruck, later explained how Avery's suggestion that genes were made of DNA left him nonplussed. "You really did not know what to do with it." Other geneticists were not really that interested in the chemistry side of the issue. As Salvador Luria later said: "I don't think we attached much importance to whether the gene was protein or nucleic acid. The important thing for us was that the gene had the characteristics it had." As one of his colleagues pointed out, their focus was "simply elsewhere."

In some instances, yet other issues were involved. There were scientists who had strong personal reasons for rejecting the DNA hypothesis. Several prominent biochemists had built their careers emphasizing the central role of proteins and were not about to give this up in order to support Avery's findings. One prominent individual actually sowed the seeds of doubt among his colleagues through articles, lectures and at conferences. His criticisms of Avery's findings seriously undermined the confidence his colleagues had—particularly those who were not biochemists—in Avery's claims.

In 1951, at a Cold Spring Harbor symposium, one of the speakers, Harriet Ephrussi-Taylor, surveyed the seven years that had passed since Avery had published his earthshaking paper. She was downbeat, pointing out that despite the interest that Avery's findings had aroused "it is surprising that so few scientists are at presently working in this field." (Cobb 2016; Holdrege 1996; Morange 2020; Mukherjee 2015)

Mindset Makes All the Difference

It is important for the students to see—as they have in other contexts in their high school biology classes—that the way you look at something influences greatly what you see. Already in 9th grade they learned by looking at ambiguous figures that "there is more to seeing than meets the eye" (N.R. Hanson). Once again—this time in the field of genetics—they recognize how the mindset that scientists bring to an issue influences greatly how they end up seeing (interpreting) and working with it.

On the other side of this fundamental phenomenon, however, is the possibility that unexpected findings will spark new ways of looking, will awaken new questions and will help individuals to reframe their and other's findings in innovative ways. Something like this also took place in the years that followed the publication of Avery's research in 1944. Soon after Avery's results were available, William Astbury—a pioneer in the use of X-rays to study crystal structures—declared them to be one of the most remarkable discoveries of all time, and proclaimed: "I wish I had a thousand hands and labs with which to get down to the problem of proteins and nucleic acids." In this spirit, Chargaff began, as mentioned above, to reexamine the chemical makeup of DNA to see if it was really as "boring" as Levene and others thought it to be. At the Pasteur Institute in Paris, Andre Boivin was inspired to undertake investigations similar to Avery's on a completely different bacterium, and came to similar results. Pioneering cell biologist Daniel Mazia began clarifying the criteria that had to be met by the vehicle of heredity, and was soon convinced that proteins did not meet all those criteria, thus opening the door even further for nucleic acids (DNA). In these and other ways, those who were excited by Avery's findings began to discover new avenues for getting to the bottom of the "genes as substance" question. (Cobb 2016, Morange 2020)

Hershey and Chase: A Turning Point

What many consider to be the turning point for this question came in 1952. It happened at the Cold Spring Harbor Laboratory on Long Island, where Alfred Hershey and Martha Chase began a series of experiments with bacteriophages, a type of virus that parasitizes bacteria.[79] When Hershey and Chase started their bacteriophage ("phage") research, it was already known that when a phage infects bacteria it begins by sitting on the bacteria's outer surface. Unclear, however, was how the phage proceeded from there. How did it take over the bacteria's metabolic system and produce hundreds of phages in the bacteria's interior, after which the bacteria would break open and release the phages—ready to consume more bacteria—into the environment?

The phage was known to consist of two basic materials: protein and DNA. By growing the phages and bacteria in a radioactive medium, it was found that the DNA would take up radioactive phosphorus, while the proteins would take up radioactive sulfur. This made it possible to track the DNA and protein after the phage took control of the bacteria and transformed it into phages. The radioactive tracers showed that the sulfur-tagged protein remained outside the bacteria like an outer shell, while the phosphorus-tagged DNA was injected into the interior. Since only the DNA entered the bacteria and 20 minutes later the cell would burst open, releasing hundreds of phages into the environment, it seemed clear that the DNA must contain the genetic material capable of creating phages out of the bacterial substance. Furthermore, the new phages contained a great deal of radioactive marked DNA. (Cobb 2016; Moore 2003; Pfeiffer 1969; Rappoport 2017)

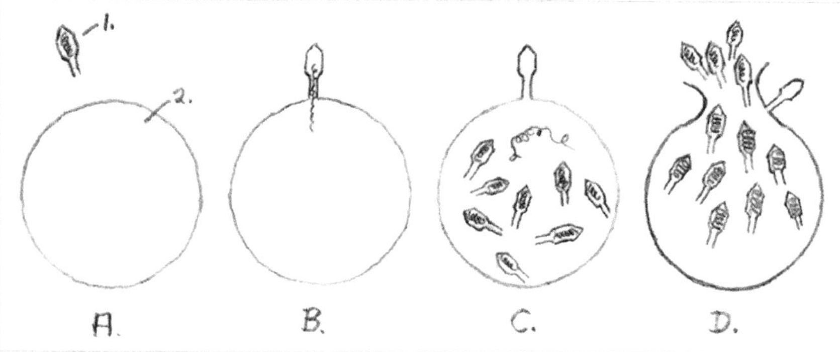

Fig. 2.14 Bacteriophage (phage) infecting a bacterium. 1. Phage. 2. Bacterium. **A.** Phage approaching bacterium. **B.** Phage docks on bacterium and injects its DNA into the cell. The phage's protein sheath remains outside. **C.** Phage DNA reproduces hundreds of phages using bacterial substance. **D.** The bacterium bursts open, releasing phages that infect further bacteria.

79 Evelyn Fox Keller (2001) maintains, for example, that it was only after the Hershey and Chase's experiments in 1952 "that most biologists were won over to the view that the genetic material was made up of DNA." That said, not all historians of 20th century biology (Morange, for example) think that Hershey and Chase's experiments should receive as much credit as they typically do in biology textbooks. Nonetheless, I emphasize their findings because they are easy to describe and easier for the students to follow than much of the related research being done at that time.

Watson and Crick, Wilkins and Franklin – The Double Helix

All of this has prepared the way for the discovery of the double helix, which is one of the best known "dramas" in the history of science.[80] It is valuable to give the students a taste of these dramatic events because they provide a rich, many-faceted context for something that they have heard about here and there for years as "the explanation" of why people are the way they are. "It's in your DNA" is one of those all-pervasive clichés similar to "the heart is a pump" and "the brain thinks." Although the story of its discovery does not confirm or deny the significance and role of DNA in living beings, it lays the groundwork for further investigation into that central question.

"Like nothing else in literature," C.P. Snow observed, "it gives one the feel of how creative science really happens. It opens a new world for the general, non-scientific reader. "

Prominent historian of modern biology, Matthew Cobb (2016, p. 90) characterizes these events in this way:

> *The story is of fundamental scientific importance and shows how science is an intensely human, collaborative and competitive enterprise in which luck, ambition and personality can play a central role.*

In reference to his own narrative of what took place on the way to discovering the structure of DNA, Watson notes:

> *As I hope this book will show, science seldom proceeds in the straight-forward logical manner imagined by outsiders. Instead, its steps forward (and sometimes backward) are often very human events in which personalities and cultural traditions play major roles.* (Watson 1968, p.13)

Before beginning the actual tale, a short depiction of the five main characters can help prepare the way.

James Watson (∗1928) grew up in Chicago. He was such a rapid learner that as a 15-year-old he was able to enroll at the University of Chicago to study zoology.

80 One of the best places to learn about it—although it is certainly a one-sided depiction in some regards—is James Watson's own narrative: "The Double Helix: a personal account of the discovery of DNA" (1968).

His great interest was birds. He had a strong antipathy toward chemistry and—despite being a science major—was able to avoid all classes in physics and chemistry that went beyond the introductory level. Graduating at the age of 19, he went on to the University of Indiana to study with the famous microbiologist Salvador Luria, who was one of the founders of the "phage group" with Alfred Hershey and Max Delbrück. Despite his weak background in and his aversion to chemistry,[81] Watson did his doctoral dissertation on phages. After graduation, he received a fellowship to continue his studies in Europe, which eventually led him to Great Britain and the famous Cavendish Laboratory at Cambridge University.

Maurice Wilkins (1916–2004) Watson was drawn to the X-ray diffraction work done at the Cavendish Laboratory after hearing a lecture in Naples by Maurice Wilkins, in which Wilkins described his research into the structure of DNA using X-ray diffraction methods. Wilkins (*1916) was a PhD in physics, who had studied at Cambridge and worked during World War II in California on the Manhattan Project that created the first atomic bomb. Shaken and disillusioned by the horror of Nagasaki and Hiroshima, Wilkins turned away from physics after the war and became interested in molecular biology. On returning to England, he took a position at Kings College in London, where he used X-ray diffraction techniques to investigate the structure of DNA.

Frances Crick (1916–2004) was raised in a small village near the English town of Northampton, where Crick's father and uncle ran the family's boot and shoe factory. After an undergraduate degree in physics, he began his PhD studies in physics at the University College London, which were interrupted by World War II. During the war he worked on the design of magnetic and acoustic mines, and was instrumental in designing a new mine that could not be detected by German minesweepers. After the war, Crick, like Wilkins, turned from physics to biology—despite having a very limited background in that area. He became a member of the research team at Cavendish Laboratory, where he worked on the description of protein structures using X-ray diffraction.

Rosalind Franklin (1920–1958) grew up in an affluent London suburb with nannies and servants, and with frequent holiday trips to Norway and the south of France. At age 17 she had a seat at the coronation of King George VI. Since she was an excellent French speaker and admirer of the French culture and life style, after receiving her PhD in chemistry from Cambridge she found her dream job in

81 In his writings, Watson frequently refers to his avoidance of chemistry.

Paris. She joined a 20-strong group there that used X-ray diffraction to study the structures of minerals. She quickly became an expert at these techniques and was soon treated as an intellectual equal by the director of the program. Being such a strong Francophile, it was a surprise to many when she decided to return to the—in her eyes—dullness of London, where she was hired to assist Maurice Wilkins as an X-ray crystallographer at Kings College in London.

Linus Pauling (1901–1994) was, in contrast to the other protagonists in this drama, already famous by the time the events we will be describing took place. In fact, at that time he was already considered by many to be the greatest chemist of the 20th century. Pauling received his PhD in chemistry from Caltech, where he also studied X-ray crystallography. Already a full professor at Caltech by the age of 29, he was far away from the action in Cambridge and London, which—as we shall see—made him a particularly dangerous competitor for those in Great Britain. His seminal paper in 1951 on the three-dimensional structure of proteins (alpha helix) was to have a huge effect on the research community we will be describing here. (It led—together with other pioneering research he had accomplished—to a Nobel Prize for Pauling in 1954.)

After orienting the students briefly about the backgrounds of our five main actors, we can begin to follow the plot as it unfolds. (Crick 1988; Cobb 2016; Morange 2020; Mukherjee 2016; Strathern 1997; Watson 1968; Williams 2019)

In Cambridge: Lots of Talk, Little Data

When Watson and Crick first met at the Cavendish Laboratory they clicked immediately. (As one biologist put it: They were "like two complementary bases snapped together by hydrogen bonding.")[82] Watson could not believe how much fun it was to interact with Crick, who "talked louder and faster than everybody else." He was a man who was constantly dreaming up and throwing out new ideas. His expansive personality and booming voice was said to often "send his coworkers running for cover and a bottle of aspirin." Crick, too, was electrified by Watson. Twelve years his junior, Watson was both brash and arrogant, with a mind as sharp as Crick's. They talked nonstop at the lab in Cambridge, at the local pub, while punting[83] on the River Cam, or over dinner at Crick's home. Their conversations were so loud and energetic that Sir William Bragg, the head of the laboratory, put

82 If we share this little metaphor with the students, we can point out to them that they will only understand the metaphor's true significance and why it "fits" this situation when we have reached the end of the story.

83 Punts are narrow flat-bottomed boats propelled by a long pole.

Fig. 2.15 Crick and Watson chatting.
(A. Barrington Brown/Science Photo Library)

them in a room by themselves so they "could talk to each other without disturbing anyone else."

Despite the magical chemistry between them, at first glance Watson and Crick did not seem to have much potential as a research team. Crick had only two years of biology under his belt and Watson had little chemistry and no X-ray diffraction experience. As a result, their original conversations were wide ranging: genetics and phages from Watson, mathematics, physics and protein chemistry from Crick. But soon enough, their interests began to converge on a single topic that had nothing to do with the protein structures that both of them were supposed to be pursuing at the Cavendish lab. (Watson, for example, was attempting—but failing—to produce crystals of myoglobin from the heart of a retired race horse.) The topic that united them was their growing conviction that the key to understanding genes lay not in proteins, but in the unraveling of the structure of DNA. Standing in their way of pursuing this question, however, was the not-minor detail that they lacked access to X-ray diffraction images of DNA, which they would need—at a minimum—in order to have a starting point for developing ideas about DNA's structure. (Cobb 2015; Williams 2019)

Fig. 2.16 Maurice Wilkins. (Public domain, von Website der National Institutes of Health)

Fig. 2.17 Rosalind Franklin. (Cold Spring Harbor Laboratory/Wikimedia Commons)

In London: Poor Communication Hinders Effective Collaboration

This brings us to the very different dynamic that existed between Maurice Wilkins and Rosalind Franklin at King's College in London. Wilkins' pet project at King's was the attempt to solve the three-dimensional structure of DNA with the help of X-ray diffraction and X-ray crystallography images. At Wilkins request, in late 1950 a new researcher—Rosalind Franklin—was recruited because her expertise in X-ray crystallography was needed to supplement Wilkins' skill set.

Early in 1951, when Franklin arrived, Wilkins was away on holiday and then missed the initial meeting with her. What Wilkins did not know (and only learned many years later) was that the Director of the Biophysics Research Unit at King's College,[84] Sir John Randall, had said in his letter of appointment to Franklin that: "as far as the experimental X-ray effort [on DNA] is concerned, there will be at the moment only yourself and Gosling." (Up until that point, Gosling had been a PhD student under Wilkins. Now, according to Randall, he would be assisting Franklin.) Since Randall had not informed Wilkins of his decision to give Franklin sole responsibility for the DNA effort, when Wilkins returned from his vacation he was under the assumption that Franklin would be supporting him on the DNA project that he had started. But on the basis of Randall's letter, Franklin had very different expectations: that the DNA X-ray diffraction work was going to be her project—and hers alone—with Gosling assisting her.

84 Wilkins was Assistant Director.

This confusion of roles, for which Randall was clearly to blame, was confounded immediately by the extreme contrast in their personalities. Wilkins was quiet, cautious and hated arguments. Franklin, by contrast, was forceful and thrived on rough and tumble debate. As her friend Norma Sutherland remembered: "Her manner was brusque and at times confrontational—she aroused quite a lot of hostility among the people she talked to and seemed quite insensitive to this." Wilkins found her not only intimidating, but was completely bewildered by her refusal to work with him. (As was said, Wilkins had no idea of the way Randall had described Franklin's new position to her.)

Franklin, from her perspective, was put off by Wilkins' reticent nature and also surprised to learn that his expertise in X-ray diffraction was not at her level. (Which was, of course, Wilkins' reason for wanting her assistance in the first place.[85]) All in all, Randall had set up "a marriage made in hell" by creating in both players very different expectations of the roles they were to play without informing them of this. A situation made all the more complicated by their severely contrasting personalities.

During the summer of 1951, Wilkins grew more and more convinced that the X-ray data was indicating that DNA had a helical structure like a spiral staircase. However, when he first presented this idea to Franklin she responded by telling him to stop working on X-ray data from DNA and "go back to your microscopes." Later, when Wilkins told Franklin that another member of the King's group was finding mathematical support for Wilkin's idea, she responded furiously: "How dare you interpret my data for me?" (Cobb 2015; Williams 2019)

Betting on DNA as a Helix, But Lacking Expertise and Sufficient Data

That November Wilkins visited Crick (his old friend) and Watson in Cambridge. He shared with them his findings to date, as well as the fact that he was not sure what Franklin's data showed because she was not sharing her results with him. His view that DNA was a helix was convincing to Watson and Crick, but how the structure was held together, how many helical chains it contained, etc., were complete mysteries when based only on the data he had access to. His great hope was that Franklin's findings might fill in some of the gaps. Since Franklin and

85 Many years later, after learning of Randall's letter, Wilkins stated: "Randall was very wrong to have written to Rosalind telling her that Stokes and I wished to stop our work on DNA, without consulting us. After Raymond [Gosling] and I got a clear crystalline X-ray pattern I was very eager to continue that work. ... Trying to understand 'what really happened' when a very admirable scientist [Randall] models himself on Napoleon is not easy ... [but the letter] was very damaging to her and to me." (Wilkins 2005, p. 145)

Wilkins were scheduled to talk about their research at a colloquium 10 days later at King's, Watson asked if he could attend and was given a formal invitation by Wilkins. Since Watson understood nothing about X-ray diffraction, he crammed as much as he could on the topic in the 10 days leading up to the seminar. It was not enough. Franklin had spent 5 years learning the ins and outs of X-ray diffraction, and his 10 days just couldn't get him up to (even close to) that level. Franklin's presentations were those of a professional crystallographer[86] for other crystallographers and she did not make concessions for amateurs. Watson was left overwhelmed and with very little that he could take home to share with Crick the next day. The one important takeaway, however, was that the DNA images she shared showed "evidence for a spiral structure."

This was enough for Crick, who was already leaning in that direction, and within a week the Cavendish pair had produced a model for DNA—a triple helix. Not experts in the field—they naively invited Wilkins, Gosling and Franklin down to Cambridge to view their creation. Franklin took one look at their model and dismissed it—pointing out how it did not cohere with her X-ray data. If such a structure were ever to exist, she exclaimed, it would instantly fly apart. She had already explained this at the King's seminar that Watson had attended but not understood. "As Franklin picked the model apart, molecule by molecule, it was as if she were extracting bones from their bodies.... By now Franklin was frankly exasperated at the 'adolescent blather.' The boys and their toys turned out to be a monumental waste of her time. She caught the 3:40 train home." (Mukherjee 2016, p. 152)

In short, Watson and Crick's first model-building venture ended in an embarrassing failure. It also had the significant effect that the King's group now realized that they were in a kind of competition that had degenerated into a race to see who could make the big discovery first. Moreover, they realized that their competitors in Cambridge were not only pushy and ambitious, but that they were tackling the question from a completely different perspective. Although they looked like rank amateurs now, they were determined opportunists who might find a way to get there first. In the end, the King's group came to the harsh realization that it was *their own data* that was being exploited, exploited by individuals who never bothered to do any research themselves!

86 Crystallography is a form of X-ray refraction.

When Randall (the director of the King's lab) heard about the Cambridge duo's efforts to muscle in on the pursuit of DNA's structure, he was furious and asked Bragg (the head of the Cavendish laboratory) to tell the two upstarts to stay away from DNA. Bragg, who was already fiercely irritated with Crick for a number of reasons, willingly forbid them from doing any more work on DNA. King's had been working on DNA for years; henceforth it would be exclusively their domain. (Cobb 2015; Mukherjee 2016; Watson 1968; Williams 2019)

The London Approach: Patience and Attention to Detail

This fiasco also solidified Franklin's conviction that before all the X-ray and mathematical data were firmly nailed down, building speculative models was ridiculous. Otherwise, models could be produced "until the cows come home" without knowing if they were even close to the truth. Once the data were really in, however, they would be self-explanatory and lead one to the right model.

Beyond these personality-influenced differences in approach, there also existed a more expertise-based difference in perspective. The British school of crystallography (X-ray diffraction) was the undisputed best in the world at that time. They were never in doubt that they would ultimately succeed in determining the molecular structure of DNA and proteins. But they were also in no particular hurry because they knew that with patience and meticulous attention to the details as they arose, they would come to clarity when the time was right and all the pieces fell into place. Rushing ahead without sufficiently processed data was not the way world class scientists went about their job. And besides, from their viewpoint, the discovery of DNA's structure was only one small step on the long march toward the scientific understanding the world we live in—a particularly urgent milestone it was not. (Crick 1988; Morange 2020; Mukherjee 2016; Williams 2019)

Model Building with the Data of Others

This was a very different mindset from the one alive in a growing number of geneticists, who saw the key to the genetic kingdom in understanding DNA. They saw themselves on the cusp of a monumental scientific breakthrough. Watson and Crick, as central proponents of this mindset, thought that Wilkins and Franklin were working too slowly because they did not grasp the earth-shaking implications of grasping DNA's structure. It was true, of course, that Watson and Crick had not done the fundamental research that understanding DNA would have to be based on. However, with their agile—and to arrogance tending—minds they saw an opening for a different approach called model building. Model building could enable them to put Humpty Dumpty together—even though they had never seen him with their own eyes (or with X-ray machines) and were basing their models on the hard work done by countless others (Wilkins and Franklin, in particular).

As we move into 1952, just over a year remained before what many consider "the greatest advance in 20th century biology" would be achieved. But for this to happen, many crucial pieces of the puzzle still had to be found and put in place. Although the people at King's College had been successful in putting down the Cavendish rebellion, they had little to offer in its place. Working out the structure of DNA would be a major triumph, for sure, but to be truly ground breaking, it would also need to explain how DNA is able to create an exact replica of its own structure during cell division, thereby providing each daughter cell with an exact copy of itself. (Crick 1988; Morange 2020; Mukherjee 2016; Williams 2019)

Linus Pauling Enters the Race

So who would find the solution? Due to their "ungentlemanly behavior" Watson and Crick had been disqualified by their higher-ups from participating any longer in the race, whereas the potential team of Wilkins and Franklin—as we have heard—was totally dysfunctional. Of the five-some we introduced at the outset of this section, only Linus Pauling had yet to step forward. But he was far away in California and seemed to be focused on protein, not DNA. In short, things looked pretty dire for the big discovery. But then, things began to shift.

As spring emerged, Wilkin's luck finally appeared to be changing. He discovered that the sperm sacks of squids lend themselves better to X-ray diffraction studies than anything he had come across before. Using a new camera that he had built, Wilkins found quite emphatic "X" patterns in the sperm DNA of squids that pointed strongly to a helical structure. With his mood buoyed by a romantic vacation to the Alps of Austria and Switzerland, he wrote a four-page letter to Crick while taking the train from Innsbruck to Zürich. Among other things he noted: "I have got much better X-ray pictures of squid sperm, with a whole series of helical lines." He included sketches of how the classic "X" pattern appeared and continued: "I think the picture holds a key not yet recognized which will then more or less guide us directly to the model.... If the helices are right, we will hit on the explanation soon." He went on to say that he saw the issues that had arisen earlier between the two labs as a thing of the past and that he looked forward to "discussing all our latest ideas and results with you again," and added: "Why don't you come and have lunch with me when you're next in town." (Williams 2019, pp. 316–317)

At this point, without realizing it, Linus Pauling began to play an influential role in the methodology and motivation of Watson and Crick. Early on in their search for DNA's structure, they had noticed Pauling's approach to understanding the alpha protein. He did not get his answers "by only staring at X-ray pictures." His primary working tool was a set of modular models reminiscent of the tinker-toys found in some early childhood classrooms. Watson and Crick now saw no reason why this would not work for DNA as well. They were soon playing with their own models, with pieces representing four nitrogen bases that would link to sugar-phosphate backbones (These were—thanks to the research of Chargaff and others—known components of DNA at that time). The X-ray pictures helped them narrow down the possible three-dimensional structures that could be composed of the known components. The clearer the X-ray images, the tighter the parameters that they could work within, but in the end, it would be the models that would bring the solution!

Early on, Pauling had shown them a new way of working, but now, in early 1953, he lit a fire under them. As it turned out, Pauling had been quietly working on DNA without announcing it to the "competitors" in London and Cambridge. Their worst nightmare had just come true: They knew that if anyone could solve the problem, it would be Pauling. To their great relief, however, it turned out that the paper Pauling had just written,[87] which proposed a structure for DNA, contained some significant and—for a person of Pauling's expertise—surprising errors. Watson and Crick saw immediately that his structure would not work. In their jubilance, they ran off to their favorite pub, the Eagle, to celebrate (with shots of whisky that were glowing with "schadenfreude").

They soon pulled themselves together, fervently hoping that Pauling's colleagues would be so in awe of him that they would not check the details of his model right away, thus delaying the discovery of his mistakes. For they knew one thing for sure, once his errors became known, the brilliant Linus Pauling would stop at nothing until he had captured the right structure. Since the manuscript was soon to be published, it would only be a matter of weeks at the latest before the error would be uncovered and Pauling would be focusing all his energy on DNA again. If that picture couldn't motivate Watson and Crick, then nothing ever would. It was now or never! (Crick 1988; Watson 1968; Williams 2019)

87 Prior to it being published, Pauling had sent the manuscript to his son Peter. Peter now shared an office with Watson and Crick at the Cavendish labs and immediately made the manuscript available to them.

"Photograph 51" is the Game Changer for Watson and Crick

In late January, Watson went to visit Wilkins in London. Things had changed in the King's lab of late. Franklin had decided to end her work on DNA there and would soon be leaving for Birbeck College. A few days prior to Watson's visit, Franklin's assistant, Ray Gosling, had come to Wilkins and presented him with an X-ray diffraction photograph from Franklin. Gosling had told him it was his to do with as he wished. Wilkins was stunned. Taken nine months prior, it was extraordinarily clear, showing the "X" form of DNA more distinctly than anything he had ever seen. This image would go into history as Franklin's iconic "Photograph 51."

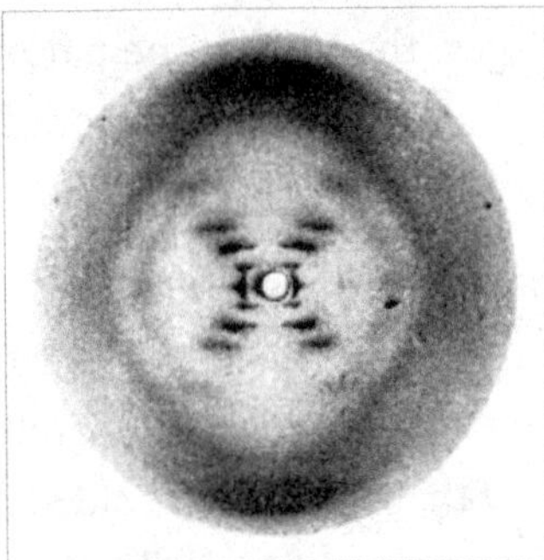

Fig. 2.18 X-ray diffraction image of the double helix structure of the DNA molecule, taken 1952 by Raymond Gosling, commonly referred to as "Photo 51," during work by Rosalind Franklin on the structure of DNA.

When Watson came to visit, Wilkins pulled the crucial photograph out of a drawer and showed it to him. Watson couldn't believe his eyes. "The instant I saw the picture my mouth fell open and my pulse started to race. The pattern was unbelievably simpler than those obtained previously.... Conceivably, after only a few minutes' calculations the number of chains in the molecule could be fixed." (Watson 1968, pp. 132–133) On the train ride home, Watson sketched what he remembered of the picture on the edge of a newspaper. By the time he arrived, he was convinced that DNA had to consist of two intertwined helical chains.

The next morning Watson and Crick rushed down to the lab and started building models in earnest. With the "Photograph 51" in mind, and knowing that Pauling was hot on their heels, they had no time to lose. "Geneticists count; biochemists clean. Watson and Crick played. They worked methodically, diligently, and carefully—but left enough room for their key strength: lightness. If they were to win this race, it would be through whimsy and intuition; they would laugh their way to DNA (Mukherjee 2016, p. 134)." (Watson 1968; Williams 2019)

The Double Helix Model Emerges

They soon concluded that the sugar[88] phosphate backbone would have to be on the outside. Playing with cardboard cutouts in the shape of the four nitrogen bases, Watson had the idea that in the interior of the helix the bases could pair up with each other, forming rungs between the sugar phosphate sides of a spiraling ladder. Since Adenine (A) and Thymine (T) were always present in nearly identical proportions, as were Cytosine (C) and Guanine (G), these pairs would form the rungs. The complementary pairs of bases—united by hydrogen bonds—could now be pictured like two opposing teeth in a zipper. Putting all of this together into a visual image, it is often represented something like this.

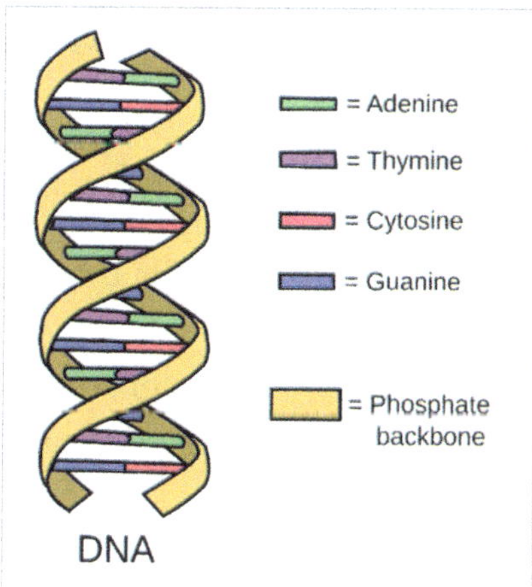

Fig. 2.19 Simple diagram of double-stranded DNA. (Forluvoft. Public domain)

Watson and Crick built their first complete DNA model in early March 1953. The Cavendish metal shop then turned the pieces into shiny metallic sections that Watson and Crick put together, part by part. Each piece had to fit and match the known molecular dimensions. In the end, it all fit like a perfectly solved puzzle and all the measurements—every angle and width, all the spaces separating the components—were nearly perfect. (Crick 1988; Mukherjee 2016; Williams 2019)

Wilkins and Franklin Confirm the Double Helix Model

The next morning, Wilkins was invited to look at the model—which was almost six feet high, even though it only represented a single turn of the helix. He liked it right away. He returned to London the next day and confirmed that his most recent X-ray data, and Franklin's too, supported a double helix. Several

88 The sugar is Deoxyribose, hence the name Deoxyribonucleic Acid (DNA).

Fig. 2.20 Watson and Crick with their DNA model.
(A. Barrington Brown/Science Photo Library)

days later Franklin came in to see the model and was also quickly convinced of its correctness. It was a beautiful solution and she recognized it immediately. When Watson and Crick published a paper describing their solution in *Nature* on April 25th, it was accompanied by one from Gosling and Franklin, which provided strong crystallographic evidence for the model, as well as a third article by Wilkins, which also corroborated the evidence with further experimental data.

Watson and Crick concluded their paper with a final, supremely *understated* line that would go down in history: "It has not escaped our notice that the specific pairing we have postulated suggests a possible copying mechanism for the genetic material."[89]

It is fruitful for the students to discuss what they think was meant by the "understated line" and why what it alluded to was so important. Breaking them into smaller groups for this often leads to lively interchanges and a healthy digestion process of the material recently covered.

After being so understated in their first paper, it was surprising to many that only a few weeks later—in May of 1953—Watson and Crick published a second, far

89 There was much discussion in the years that followed about why Watson and Crick remained so modest in the understated conclusion to their paper. Crick clarified this years later, describing it as a compromise: "I was keen that the paper should discuss the genetic implications. Jim was against it. He suffered from periodic fears that the structure might be wrong and that he had made an ass of himself. I yielded to his point of view but insisted that something be put in the paper, otherwise someone else would certainly write to make the suggestion, assuming that we had been too blind to see it. In short, it was a claim to priority." (1988, p. 66)

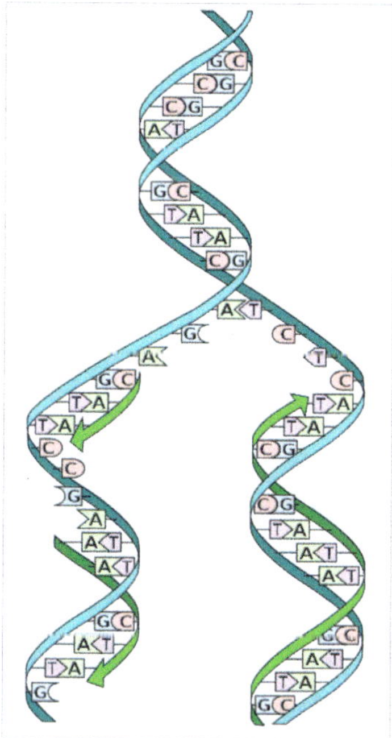

Fig. 2.21 Simple diagram of DNA replication. (Madprime. GNU Free Documentation License)

more speculative one, which expanded significantly upon the first.[90] In the second paper they explain how a stretch of DNA could duplicate itself in preparation for cell division. For this to happen, the hydrogen bonds connecting the base pairs would be broken before duplication, allowing the two strands of the Helix to separate and unwind without everything getting tangled. Thereafter each half would act "as a template for the formation onto itself of a new companion chain," thus producing two exact replicas of the original double helix, each with exactly the same sequence of bases as in the original. This is illustrated in the diagram above (Crick 1988; Mukherjee 2016; Williams 2019)

Winning the Nobel Prize

In 1962, Watson, Crick and Wilkins won the Nobel Prize for discovering the structure of DNA. Rosalind Franklin passed away from ovarian cancer in 1958 at the age of 37 and was not included in the prize. This was because, according to the Statutes of the Nobel Foundation, a Nobel Prize cannot be awarded posthumously. This tragedy reminds the students of what happened to Hilde Mangold, and they appreciate hearing a bit more about Rosalind—to learn that although some

90 As it turns out, when they sent in their draft of the initial paper, they had not yet seen the two other papers by Franklin & Gosling, and by Wilkins, whose X-ray evidence strongly supported their structure. Later, when they saw those papers, Crick easily persuaded Watson that they should write more.

experienced her as aloof and withdrawn in the laboratory (set up by Randall's original confusion-creating letter), outside King's she was apparently happy, social and enmeshed in a network of friends and family (which, it is said, included evenings out with the leader of a symphony orchestra).

After rounding off her work on DNA in 1953, her relations with both Watson and Crick grew much warmer. For example, during a busy lecture tour of America—she had become, in the meantime, an internationally known expert on viruses—she ran into Watson in Massachusetts and he offered to drive her (together with a colleague) across the United States all the way to California. Unfortunately, she had to decline, because she had already arranged to meet with Crick in Brooklyn, where he was now working. In March of 1956, after presenting at the International Crystallography Conference in Madrid, she got on so well with Crick and his wife Odile, that she joined them for a tour of Southern Spain after the meeting.

In 1956 her health began declining and after a small operation, she stayed with the Cricks at their home in Cambridge while convalescing. After bouncing back toward the end of 1957, her energy began to seriously decline. By late March 1958 she was still working half-days most of the time, but could only crawl upstairs. In mid-April 1958, on the day she had been scheduled to speak at the Faraday Society in Leeds, she passed away, three months short of her 38th birthday. In a tribute to Franklin, Wendell Stanley, a Nobel laureate in chemistry, wrote how her death cut short "a life of great promise and complete devotion to scientific research." She had been "a woman of great intelligence and wide culture, and an international courier of good will," who made "great contributions" in several fields of science. (Crick 1988; Mukherjee 2016; Williams 2019)

Although it is possible to carry with on with the story-form and describe the human dynamic that informed the path from the double helix to the genetic code—starting with the letter George Gamow sent from America to Watson and Crick —I usually move on at this point to a more general explanation of the role of DNA in the organism developed by geneticists in the next decade and a half.

Protein Synthesis: The Central Dogma and the Genetic Code

Returning to the process of DNA replication described above, we can add a bit more detail and terminology that the students will need to understand the process of protein synthesis and what is meant by the word "code."

As described earlier, in order for DNA to duplicate itself, the two stands that make up the rungs of the ladder must separate at the hydrogen bond, splitting the ladder in two as if it were a zipper opening. Each half-strand now serves as a template for complementary free-floating nucleotides[91] that combine with them, thereby forming a new, identical double helix—because guanine (G) can only combine with adenine (A), and cytosine (C) with thymine (T). During mitosis these two exact replicas of the original double helix part company and move into a daughter cell. Based on what we learned earlier in the block, enzymes are necessary for these processes to take place. The helicase enzyme is involved in the unwinding process, and another enzyme, DNA polymerase, is engaged in the joining together of the two nucleotides as they unite to form—step by step—two exact replica of the original "spiral ladder." (More on these enzymes and the exactness of the replication process soon.)

With a description of this kind, we have illustrated briefly the basic idea for which Watson, Crick and Wilkins received the Nobel Prize. The riddle of how genetic material could be preserved and passed on to each new cell within the organism, as well as to later generations, appeared to have been solved.

Between 1953 and 1966—sometimes called the golden age of genetics—basic findings about DNA came fast and furious. By the mid-60s, scientists had uncovered the central aspects of its functioning. It was seen to have two basic tasks: replication, as we have just heard, and the production of proteins. Since the latter process is extremely complex, a general overview at the outset provides the context for explaining some of the more detailed discoveries made in this field.

As the students have already learned, the name protein means "primary," which is a fitting name for the most abundant substance—next to water—in living organisms. They have also learned that proteins are made up of strings of amino acids (polypeptide chains) and that these typically range from 50 to 2000 in number. Based on the kinds of amino acids and on their sequence in the string, a huge number of different proteins become possible. To date, more than 100,000 have been identified. Our question in this context is: How do they get strung together in a specific order that creates this or that protein? In other words, how does the limited number of variables in our double helix—four different nitrogen bases arranged in various sequences—enable the creation of over 100,000 different proteins? This is very hard for the students imagine, as it was for most geneticists in the first half of the 20th century. With this question in mind, we begin with the overview.

91 The basic threefold-unit of the ladder consists of one sugar-phosphate plus one base, and is called a "nucleotide."

Transcription. Whereas DNA is found in the nucleus of the cell, another nucleic acid, RNA, is found primarily in the cytoplasm outside the nucleus. The two are much alike except that RNA is single-stranded (one handrail), its sugar is ribose instead of deoxyribose, and of the four bases, it substitutes uracil (U) for thymine (T). In the first stage of protein synthesis, called "transcription," the DNA opens up in waves along the helix and the order of its bases is used as a template by a strand of RNA—called messenger RNA (mRNA)—for creating a complementary strand of mRNA, with the only difference being that U replaces T.

Translation. In the second phase of protein synthesis, the strand of mRNA moves outside the nucleus into a ribosome located in the extra-nuclear cytoplasm. Each sequence of three bases, called a "triplet," forms a "codon" (code word) for a specific amino acid. With the help of enzymes, amino acids are brought to the ribosome by a second form of RNA, transfer RNA (trRNA). There, based on the sequence of codons found in the mRNA (that were copied from the DNA), a corresponding succession of amino acids are linked together to form a growing polypeptide chain, which results lastly in a specific protein. In a shorthand formulation of what he called the "central dogma," Crick summarized this process: "DNA makes RNA, RNA makes proteins, proteins make us." (The "central dogma"

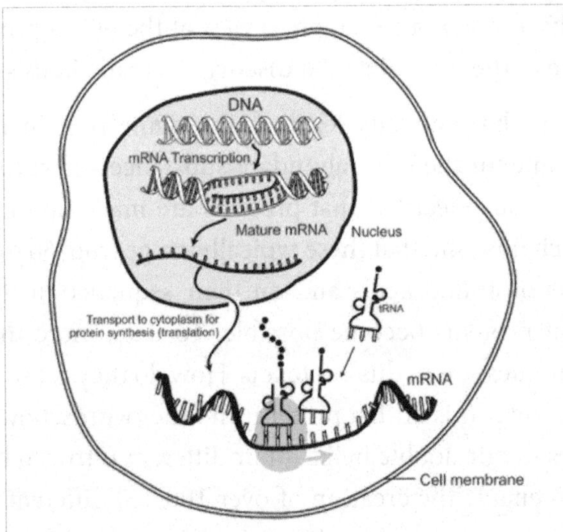

Fig. 2.22 Illustration of transcription and translation. In the nucleus, DNA is transcribed to mRNA. After being exported to the cytoplasm, the mRNA molecule is bound by a ribosome and is translated using tRNAs, which match amino acids to codons in the mRNA. (National Human Genome Research Institute, NIH. Public domain)

also emphasizes that this sequence cannot take place in the reverse order.) This entire process can be illustrated with a board drawing something like Fig. 2.22.

What has just been described illustrates what is known as the "genetic code." The sequence of bases in the DNA is called a "triplet code" because each group of three bases (a codon) codes for an amino acid that has been "transcribed" by mRNA and then "translated" in a ribosome into an actual amino acid with the help of trRNA. After such a significant discovery, we ask the students where they think things might go from there. Is the work of geneticists now complete and they will soon be unemployed? (Hickman et al 2001; Krogh 2010; Raven 2016)

The Human Genome Project (HGP)

Based on the understanding that protein synthesis is determined by the sequence of base pairs in DNA, and on progress in the area of DNA-assessing technologies, in the mid-1980s an initiative arose to analyze the structure of the entire human genome. That meant determining the sequence of bases contained in all the DNA that is divided up in the 23 chromosome pairs found in our body cells. By 1990 the necessary funding for the project had been found and the Human Genome Project (HGP) began.[92] It was coordinated by the US Department of Energy (DOE) and the National Institute of Health (NIH).[93]

Central among the goals of the HGP were to

- Identify the sequence of the base pairs in the human genome
- Identify how these sequences manifest as specific genes
- Determine the genetic profiles of countless hereditary diseases

The project was begun with a great deal of hype and grandiose predictions. For example, Frances Collins, head of the NIH's role in the project, made forecasts such as the following (Michio 1997, p. 143):

- *By the year 2000 scientists will have deciphered the genetic codes for 20 to 50 hereditary diseases which have caused untold suffering since the dawn of humanity, including cystic fibrosis, muscular dystrophy, sickle-cell anemia, Tay-Sachs disease, hemophilia, and Huntington's chorea.*
- *Prior to 2006 the 100,000 or so genes*[94] *that make up the human genome will have been deciphered by the Human Genome Project, which will open up secrets locked for millions of years in our genes.*
- *By 2010 the genetic profiles of hereditary disease will balloon to approximately 2,000–5,000, giving us an almost complete understanding of the genetic basis of these ancient diseases.*
- *By the year 2010 you will be able to have your own report card printed out of your individual risks for future disease based on the genes you have inherited.*

92 There is a much more complicated "backstory" to how the HGP unfolded, one that involves Craig Venter and his firm Calera, that I usually do not go into here.

93 Its international analogue was the Human Genome Organization (HUGO).

94 At that time this was a common prediction for the number of genes that would make up the genome.

- *By 2020 or 2030, you'll be able to go to a drugstore and get your own DNA sequence on a CD, which you can then analyze at home on your Macintosh.*

In June 2000, a ceremony was orchestrated in the White House, where President Clinton, together with the two central figures in the genome mapping efforts—Frances Collins and Craig Venter—announced that the first draft sequence of the genome had been completed. Clinton had this to say:

> *Nearly two centuries ago, in this room, on this floor, Thomas Jefferson and a trusted aide spread out a magnificent map, a map Jefferson had long prayed he would get to see in his lifetime.... It was a map that defined the contours and forever expanded the frontiers of our continent and our imagination. Today the world is joining us here in the East Room to behold the map of even greater significance. We are here to celebrate the completion of the first survey of the entire human genome. Without a doubt, this is the most important, most wondrous map ever produced by humankind."* (Mukherjee 2016, p. 318)[95]

Completed in 2003, some of the key findings of the project were:

- The human genome consists of over three billion base pairs (~3,088,286,000)
- Genes were to be defined as specific sequences of bases that "encode" protein synthesis.
- Genes comprise only about 2% of the human genome! The remainder consists of non-coding regions.
- The human genome is estimated to contain only 20,000–25,000 genes. (Far fewer than most estimates prior to the HGP, which normally ranged from 80,000 to 140,000 thousand. This is all the more surprising when we learn that a tiny roundworm has roughly the same number of genes as we, and that a water flea has 39,000, a pea aphid 34,000. As far as

95 Although it is important not to bring a cynical tone into the discussion, it can be pointed out to the students that one of the driving forces behind the HGP was the biotech industry. It had been clear since the first discoveries in molecular biology that "genetic engineering" had an immense potential for producing private profit. As Harvard's Richard Lewontin—one of the 20th century's most prominent biologists—observed, the biotech industry was firmly behind the HGP: "No prominent molecular biologist of my acquaintance is without a financial stake in the biotechnology business. As a result, serious conflicts of interest have emerged in universities and in government service." Even James Watson's resignation from his position as the first head of the NIH Human Genome Office in 1992 is linked to such issues. (Gruber 2013; Lewontin 2000; Roberts 1992)

plants are concerned, our genome has 12,000 fewer genes than corn, and 25,000 fewer that rice or wheat.)

Many more details could be described, of course, but in order to give the students a sense for the amount of data involved here, a concrete comparison such as the following makes a big impression on them. If the base sequence of the human genome (ATGTCAGTC etc.) for one human cell were published in 500-page volumes with standard size font, then it would take 400 such volumes to record it all. Even more striking is the following: If we were to read the 3 billion bases in the genome out loud (24/7, without sleeping) at the rapid rate of five per second (AATGC)—which would be 300 bases per minute, 18,000 per hour, 432,000 per day, and 157,680,000 per year—it would take us 19 years (sans sleep) to read out the entire base sequence in one human cell!

Beyond the details revealed through the HGP, it is important to note the way in which metaphors were used in reference to the human genome. Metaphors help us, of course, when we are trying to grasp something unfamiliar or very complicated by relating it to something familiar or uncomplicated. DNA and the human genome were commonly likened to a computer program that encodes the development of the organism and the way it responds to the environment. Just as a computer is described as "reading" and executing the instructions contained in its program, the body was frequently characterized as reading and executing the instructions present in the genome. From this perspective, the body is essentially viewed as "only hardware" and the genome is the boss.[96]

Although the "gene-as-text" metaphor—with DNA sequences referred to as a code, a text, a program, etc.—is common parlance, and has provided an obvious rhetorical bridge over the gap between physical substance and organic forms and processes, it has also raised questions about its actual soundness. Matthew Cobb, in his book "Life's Greatest Secret — The Race to Crack the Genetic Code," characterizes these issues in the following way:

> *The complex linguistic and computational metaphors wrapped up in the seemingly simple idea of a genetic code frame our ideas about heredity. But a "frame" means two things—it both enables and limits how we think.... Metaphors and analogies carry a risk. It is easy to forget that a particular term is a figure of speech, a way of viewing a given phenomenon, rather*

96 A less frequently used, more literary image for the dominance of the genome goes something like this: If life were a play, DNA would be the playwright and proteins the actors. Together they are responsible for all of life's essential properties. (Rothman 2002)

> *than being literally true. A gene is like a computer program, but it is not a program and does not function according to the same rules, even though it may be usefully understood in this way. Organisms are not machines, even if they work on physical principles and share some features with devices we have invented. The genetic code is not literally a code and it is not a language.... Understanding the power and limits of such metaphors will help us prepare for the breakthroughs of tomorrow, when we will reinterpret what we know and discover what we have yet to imagine. New technological and scientific achievements will provide us with new metaphors, new ways of understanding how life works.* (Cobb 2015, pp. 312–313)

With these reflections in the background, we can proceed to other issues that arose as the HGP moved ahead and garnered more and more headlines in public media. (Cobb 2015; Cohen, Atlan & Etoni 2016; Michio 1997; Mukherjee 2016)

Cracks in the Central Dogma Gradually Become Crevasses

Soon after the double helix and the "central dogma" had become headline material and before the HGP was born, cracks in the veneer of DNA self-replication and DNA→RNA→Protein began to appear. On the one hand, it was evident that viewing DNA as if it were an independent entity was a high level abstraction. In the words of biologist Philippe Matile, professor at the Swiss Federal Institute of Technology (ETH Zurich):

> *The structural organization of the cell is the unavoidable foundation of metabolic processes; it is the prerequisite for the existence of DNA.... In reality, DNA does not stand at the beginning, but rather in the middle of a cyclical process in which the entire cell metabolism and the entire structural organization of the cell are included. Indeed, DNA plays an important role in the metabolic processes, but it is, on the other hand, a product of those processes. To speak of a "self-duplicating activity" by DNA makes no sense.* (Matile 1973, transl. MH)

Once it became clear that DNA must be seen as embedded in and influenced by a larger whole—the cell—and that the cell, in turn, can only be understood in the context of the organs and organism of which it is a part, other ways of speaking about DNA's role in the whole became viable. Harvard's Ruth Hubbard, for example, offers a different kind of metaphor to illustrate that role:

> *The language that geneticists use often carries considerable ideological baggage. Molecular biologists, as well as the press, use verbs like "control," "program," or "determine" when speaking about what genes or DNA do. They are all inappropriate because they assign far too active a role to DNA. The fact is that DNA doesn't "do" anything: It is a remarkably inert molecule. It just sits in our cells and waits for other molecules to interact with it.*
>
> *In a way, the DNA in our cells is like a cookbook. We need a cookbook if we want to make a complex dish, but it does not make the dish, nor can it determine which dish to make or whether the dish will come out right. The cook and the ingredients will determine whether and how a recipe is used, whether we end up eating soup or cake, and how the food tastes. Cookery is also an apt metaphor because it introduces an element of adaptability and flexibility. A good cook can deviate from the recipe and fudge the outcome.... Moreover, if cells and organisms are the cooks in this metaphor, many ingredients, among them genes and environmental factors, combine to produce a "dish" that could not have been predicted by looking at the ingredients separately.* (Hubbard & Wald 1993, pp. 11–12)

Cell Differentiation. One paradox that helped nudge molecular biologists toward a less narrow understanding of genetics and the role of DNA relates well with the cookbook idea. This is the difficulty that cell differentiation presents to an exclusively DNA-based explanation of heredity and development. With few exceptions, every different type of cell in the human body contains the same chromosomes and the same DNA sequence as the original, single-celled zygote. Yet somehow this zygote manages to differentiate into every manner of tissue — liver, skin, muscle, brain, blood, bone, etc. As we learned earlier in the block in connection with our cell differentiation diagram, during morphogenesis, a totipotent zygote http://en.wikipedia.org/wiki/Stem_cells) gives rise to various pluripotent cell lines in the embryo, which in turn lead to multipotent cells, before we come to the fully differentiated cells at the end of the line. If genes determine the form and substance of the organism, how is it that such radically different cellular architectures can result from the same genes? What directs genes so that they can produce the intricately sculpted and complexly differentiated form of a human being? What determines when and where the roughly 200 body-cell recipes are engaged, thereby causing cells to differentiate in this or that direction during embryonic development—and how is this done? (The riddle becomes even greater when we think back to Hadorn's regeneration experiments, where cells de-differentiate, re-differentiate and even trans-differentiate!)

Regulator Proteins. Already in the early 1960s, the way in which other factors are able to influence DNA and protein synthesis began to surface. The biochemical and genetic analyses done by Francois Jacob and Jacques Monod at that time uncovered what became known gene regulation.[97] They found that gene expression can be switched on or off by "regulator proteins." "Repressor" proteins can prevent RNA polymerase from transcribing a gene, thereby turning a gene OFF. An "inducer" protein, on the other hand, releases the "repressor" and the gene is now turned ON again and can be expressed with the help of RNA polymerase. With the concept of "regulator proteins," the door had now been opened for the discovery of a whole range of agents that influence genomic activity. (Hubbard 1993; Jacob 1998; Fox-Keller 2002; Shapiro 2011)

In the years since Jacob and Monod introduced this concept, a great number of other regulatory elements have been identified. The pedagogical challenge is finding a way to give the students a sense for the "new players" in the genetic landscape without overwhelming them with details that are far removed from the phenotypic reality that they live in. There are many possibilities. I will take a middle road here that can easily be expanded upon or simplified, depending on the class and time available.

Split Genes. In the 1970s an even bigger challenge for the "central dogma" presented itself with the discovery of "split genes." Most genes in higher organisms are not continuous, but are composed of "coding" segments of DNA called "exons," with long noncoding sections in between called "introns." In these instances, the original copy of the DNA sequence made by mRNA must be processed before protein synthesis can begin. What will need to happen, we ask the students? Obviously, the intron that lies between the exons must be removed so that correct sequence of bases is created that will synthesize a particular protein. This is done by special protein/RNA complexes called "spliceosomes" that cut out the introns and then splice the remaining exons together to form a continuous coding "transcript." (Cohen, Atlan & Efroni 2016; Fox Keller 2002)

97 Although their initial research referenced prokaryotes, analogous principles were later found in eukaryotes, as well.

Alternative Splicing. Once this complex split-gene splicing process was discovered and sorted out, an even bigger surprise appeared: alternative splicing. Although exons can be connected in their original sequence, they can also be spliced together in alternative ways, thus creating mRNA transcripts that are different from the original and that lead to the synthesis of new and different proteins. As it turns out, about a third of eukaryotic genes are routinely subjected to this kind of variable reading! The result is that a single DNA sequence can give rise to more than one species of protein. These alternatives are not determined by the DNA sequence itself, but by enzymes and proteins that influence the splicing.

They, in turn, are responding to the circumstances and stage of development of the organism itself. This kind of flexibility—cutting and pasting different pieces of the original "transcript" to form new, context relevant templates—is particularly important in developing organisms. The number of different proteins associated with a particular gene reaches into the hundreds in some organisms. It is estimated that approximately 35% of human genes produce alternatively spliced RNAs, which makes it possible for a limited array of genes to create a much larger assortment of proteins.

One classic example discovered in the 1990s involves a gene active in the hair cells of the inner ear of a chick. It encodes a protein that plays a role in determining the resonance frequency of the cell. Variations in the sequence of that protein alter it in order to tune the chick's ear to different sound frequencies. 576 possible splicing variants of that gene have been identified.[98] (Black 1998; Cohen, Atlan & Efroni 2016; Fox Keller 2002; Gilbert 2010)

Conformation. Other problems arose for the central dogma when it became clear that the synthesis of a protein is not the end of the story. After being synthesized, the function of a protein can be further regulated depending on its context. The way a protein functions depends greatly on how its chain of amino acids is folded. Known as "conformation," the folding of a protein is determined dynamically in response to interaction with other proteins in its environment, as well as other factors such as pH. Different conformations expose different sections of the protein and allow for different interactions.[99] (Cohen, Atlan & Efroni 2016; Gilbert 2010; Ha & Lo 2012; Orellana 2019.)

98 The record holder appears to be Drosophila, whose Dscam gene can produce 38,016 different types of proteins by alternative RNA splicing. (Gilbert 2010)

99 A lighthearted image I sometimes use to illustrate this: If you turn a home run hitting baseball star around while at bat so that s/he is facing the catcher, there will be no more home runs from that batter.

PTMs. Moreover, proteins encoded by a gene can, and often do, undergo significant "post-translational modifications (PTMs) such as glycosylation and phosphorylation that alter the activity of the protein, allowing it to carry out further significant functions within the organism.[100] (Prabakaran et al. 2012; Raven et al. 2016)

We could keep going with more examples of gene regulation, but overwhelming the students with too much "phenotype-distant" information of this sort can easily lead to "academic indigestion" that makes learning a chore rather than a catalyst for new insights. Nonetheless, one additional regulatory factor is hard to leave out. So, if the students still seem hungry for more, telling them about the elaborate system that exists to keep DNA replication error-free will amaze them.

Proofreading and Repair. Although one of the key selling points of the double helix was that its structure would allow for easy replication, it turns out that additional DNA proofreading and repair systems are absolutely necessary to insure accurate replication. If only chemical processes were involved, the replication accuracy would be approximately 1 error for every 300 base pairs. That would mean that by every cell division approximately 10 million nucleotides would be falsely replicated. However, during the polymerization process the enzyme DNA polymerase (that enables the adding of nucleotides) detects false matches and replaces them immediately. This increases the accuracy to one mistake for every 100,000 nucleotides. Although this would be very impressive for any human-made manufacturing process, it is still not enough to maintain hereditary continuity over time.

A further step is the so-called "proofreading" process that senses distortions in the growing double helix caused by mismatches. It then halts the polymerization process, removes the incorrect base, and replaces it. This process increases replication accuracy 100- to 1,000-fold. An additional stage of proofreading called "mismatch repair" utilizes three different proteins that are dedicated to post-replication inspection and repair. This highly efficient system increases the precision of replication to one mistake per 10 billion base pairs. As is more than obvious, this amazing multiphase process is not done by the DNA. Without such an elaborate system of monitoring, proofreading and repair, which is carried out by various enzymes, the replicatory process would be too sloppy and accumulate

100 Protein phosphorylation, for example, plays an important role in the coordination of several biological processes, including the regulation of metabolism, inflammation, and apoptosis.

too many errors to allow for the hereditary stability that we observe in humans and other organisms. (Shapiro 2011; Wirz 2008)

So once again, we see how it is not the structure of the DNA itself that makes such processes possible—although that structure is a necessary prerequisite—but rather its embeddedness in the astounding menu of capacities found in the organism as a whole.

The HGP in Light of a Crumbling Central Dogma

After all that the students have just learned, it is time to step back and ask: Well, what about the expectations for the HGP that we just heard about? What became of them? According to its advocates, knowing the genomic sequence was going to provide the instruction book that would show us how a healthy human being is constructed, building block by building block. Very soon we were to have in hand a gene for obesity, a gene for cancer, a gene for cystic fibrosis, etc. etc.

And yet—as The Nature Institute's Stephen Talbott observed in 2010—"the most striking thing about the genomic revolution is that the revolution never happened." Talbott goes on to say:

> *As minor tokens of the changing consciousness among biologists, one could cite articles from this past year in the world's two premier scientific journals, each reflecting upon the discovery of the "gene for cystic fibrosis." THE PROMISE OF A CURE: 20 YEARS AND COUNTING — so ran the headline in* Science, *followed by this slightly sarcastic gloss: "The discovery of the cystic fibrosis gene brought big hopes for gene-based medicine; although a lot has been achieved over two decades, the payoff remains just around the corner." An echo quickly came from* Nature, *without the sarcasm: ONE GENE, 20 YEARS — "When the cystic fibrosis gene was found in 1989, therapy seemed around the corner. Two decades on, biologists still have a long way to go."*
>
> *The story has been repeated for one gene after another.... But is this surprising?.... The fixity, precision, and unambiguous logical relations of the code seemed to guarantee its strictly mechanistic performance in the cell. Yet it is this fixity, this notion of a precisely characterizable march from cause to effect — and, more broadly, from gene to trait — that has lately been dissolving more and more into the fluid, dynamic exchange of living processes. Organisms, it appears, must be understood and explained at least in part from above downward, from context to subcontext, from*

> *the general laws or character of their being to the never fully independent details. In the end, we can meaningfully apprehend the lowest-level activities only so far as we recognize them to be performances of the whole organism.* (Talbott 2010)

On the other hand, the sequencing of the human genome has been very valuable for reasons quite different from the original expectations. As Evelyn Fox Keller has pointed out, it has helped "to reveal the naiveté of those hopes, and thereby to set us on a more realistic track toward understanding how organisms develop, function, and evolve." Fox Keller, as one of the 20th century's premier philosophers of science, draws the following conclusions from what the HGP actually taught us:

> *We learned that the source of genetic stability was not to be found in the structure of a static entity but that stability is itself the product of a dynamic process… that gene function needs also to be understood in dynamical terms. Because biological function inheres in the activity of proteins rather than of genes, the breakdown of the one-gene-one-protein hypothesis critically undercuts the possibility of attributing functions to the structural unit that has traditionally been taken as the gene. Yet reconceived as a functional unit (for example, the spliced and edited mRNA sequence), the gene can no longer be set above and apart from the processes that specify cellular intercellular organization. That gene is itself part and parcel of processes defined and brought into existence by the action of a complex self-regulating dynamical system in which, and for which, the inherited DNA provides the crucial and absolutely indispensable raw material, but no more than that."* (Fox Keller 2002, p. 71)

Epigenetics

Before rounding off this discussion, one more central dimension needs to be considered: the transgenerational aspect of gene regulation, which is known as "epigenetics." The Greek prefix "epi-" implies aspects that are "on top of" or "in addition to" the traditional genetic basis for inheritance. It refers to the study of changes in organisms that are caused by modifications in the way genes are expressed rather than alterations of the genetic code (the DNA sequence) itself. In this sense, it is basically a synonym for "gene regulation." The term epigenetics is often used in a narrower sense that refers in particular to gene regulation that is passed on from one generation to the next. The modifiers whose effects can be passed on to future generations involve for the most part three systems: DNA methylation, histone modifications, and non-coding RNA. In my experience, to go into detail about these three factors is too much for the students (and also unnecessary) after all that has just been discussed.[101]

Instead, I like to lead the class into a discussion of the overall dynamic of genetic activity referred to in the quotes from Talbott and Fox Keller above. One way to do this is to see what James Watson—now over half a century wiser than back in his Cavendish Laboratory days—has to say about epigenetics:

> *Epigenetics is currently an exciting area of research. We have long understood many of the ways in which genes are switched on and off, but epigenetics provides a new set of regulatory mechanisms.... Epigenetics refers to a suite of chemical modifications that occur at different times during development that govern the activation of particular genes. If you think of the genome as the orchestra, then epigenetic regulation is the conductor (or at least one of the conductors).* (Watson 2017, p. 382)

Exploring the metaphor of "genome as orchestra" can lead to very interesting discussions. It often makes sense to break up the class into small groups to see what they can come up with, and then to bring them all together to build up a living picture of what such a perspective might reveal. A somewhat analytical group, for example, will break the metaphor down into a number of components—setting (acoustics), instruments, players, conductor, music, composer—that need to be

101 However, it can provide material for an extra report by a student, or group of students, who are particularly interested in this topic.

brought into relationship to each other. In most instances, a lively discussion will lead to certain fundamental and unavoidable questions: Where does the music come from? Who is the composer? Who is the conductor? Where is the music when it is not being played?

To help the students see that such questions are not just playful explorations based on a farfetched metaphor, it is can be pointed out that a prominent professor of physiology at Oxford University, Denis Noble, uses a similar comparison in trying to describe the kind of causality that is involved in genetic activity. In one of his books, *The Music of Life*, he compares the genome to an immense organ with 30,000 pipes. He points out how the music is an integrated activity of the organ, not just a series of notes, and that the music itself is not created by the organ—it did not write the Bach fugues played on it, Bach did. Moreover, the music requires an accomplished organist to make the organ perform in such a way that the music can manifest in any particular moment. Noble concludes his "musical analogy" with these pivotal questions: "If there is an organ, and some music, who is the player, and who was the composer. And is there a conductor?" (Noble 2008).

Summing Up

To round off this long depiction of the development of genetics, it is valuable to take a step back and look for overall tendencies that are characteristic of that development.

Starting with Mendel, we see certain human characteristics that are often found in pioneering scientists. He had real questions that he pursued with great dedication and tenacity over many years. He also developed a unique approach to answering those questions—one that was uncommon and little understood at that time, but which was to take center stage as genetics unfolded in the century thereafter. He stopped looking at the whole organism in its environment and focused on certain discrete traits. This allowed him to assess his findings mathematically and from there to create a thought model about imperceptible "factors" that would make sense of the mathematical ratios he had found.

This focus on discrete traits and invisible hereditary factors—in contrast to more general phenotypic characteristics and developmental phenomena—went a step further when Sutton linked the invisible factors to chromosomes. In this way the focus narrowed from the organism as a complex whole all the way down to invisible (without technological assistance) subunits of the cell's nucleus, the chromosomes.[102] Soon thereafter, the "key to the kingdom" would be sought at an even more subliminal level: In invisible subunits of the chromosomes that would be called genes!

The overall tendency just described is known as "reductionism." The focus of attention was gradually reduced (narrowed) from living, sense-perceptible organisms interacting within their environment, to sub-, sub-, sub-, subunits of the larger whole. These discrete subunits at invisible levels of existence and interpreted by thought models were now seen as the heart of the matter.

As Harvard's Ruth Hubbard (2013, p. 18) notes: "At a time when invisible atoms, electrons, and quanta were being accepted into the world of chemistry and physics, biologists had little problem accepting that heredity was also mediated by invisible material particles." So where did they go from there?

102 The nuclei are, of course, subunits of cells, which are subunits of tissues, which are subunits of organs, which are subunits of the organism as a whole which, of course, plays a particular role—is in some ways a subunit—within the ecosystem in which it resides, and that ecosystem, in turn, plays a role within the biosphere of the earth.

Starting at this level—where discrete sensory characteristics are being linked to discrete subunits of chromosomes—we find T.H. Morgan and team performing tens of thousands of experiments on fruit flies. These were meticulously recorded and then statistically analyzed. All of this led to the creation of chromosome maps. This was a huge step. From the fluidity of embryological phenomena we studied earlier in the block, Morgan and others seem to have fallen back into something like the mosaic theory of August Weismann, where the starting points for all traits "stand side by side, separate from each other like the stones of a mosaic, and develop independently, although in perfect harmony with one another into the finished organism." In this way, reductionist genetic research appeared to have developed a very ironclad theory, one that stood in great contrast to the findings of embryologists. (For Spemann's experiments—the students will remember—clearly refuted the mosaic theory at the level of phenotypic development.[103])

But, lo and behold, things were not so straight forward/ironclad as the chromosome maps from the "fly room" made them appear. Everything became more fluid, complex, and interwoven again when further research revealed that in many instances one gene can influence multiple traits (pleiotropy) and that one trait can be influenced by many genes (polygenes)![104]

With these discoveries, everything came into movement again—no ironclad mosaic after all—and since "no one had seen a gene in action or knew its material essence" (Mukherjee) a different approach—a biochemical one—gradually joined the search for the secret of heredity. Pivotal early attempts in this direction moved from the complexity of higher-level organisms like Drosophila to countless experiments on a simple bread mold (Neurospora), which led to the conclusion that enzymes might be the key substance in the physiology of genetics.[105] This viewpoint found considerable resonance because enzymes facilitate almost every chemical process in the body. Enzymes are proteins, and since it was known by then that chromosomes are composed of only two substances, proteins and DNA, proteins seem to be the obvious candidate to be the carrier of genes. Made up of strings of amino acids—usually 300 or more per protein molecule—they provide

103 This shows, too, how disconnected these disciplines had become from one another.

104 Fox Keller (2010) also points to a significant distinction that was not emphasized during the glory days of classical genetics, namely the difference between what can be called a "difference maker," and what is actually a "trait maker." It is one thing to say that a gene is a difference maker and influences differences in the way a trait manifests, such as the color of a flower. It is something very different to say that the gene is a trait maker, that it is responsible for the existence of the trait as such—for the flower in its entirety.

105 Although the scientists themselves recognized that the popular "one gene, one enzyme" moniker was an oversimplification.

the potential for unbelievable diversity and variation. DNA, as it was understood at that time, appeared very monotonous by comparison and was soon viewed as merely support scaffolding for genes. Proteins carried the real stuff, DNA was just the stuffing (Mukherjee).

But, as we have learned to see, when dealing with organisms things are often not as clearcut as they first appear. Nonetheless, the "protein is the gene carrier" perspective was soon firmly rooted and not easily shaken. So when Avery's findings appeared to indicate that DNA was the primary hereditary substance, most biochemists/geneticists just assumed there must have been errors in his results. In some instances, individuals were so committed to the central role of proteins in their research and ruminations that to consider DNA as a viable candidate would have been tantamount to "switching horses in midstream"—no way!

In spite of this overriding tendency, a few researchers were open to Avery's findings and were discovering ways to explore further in that direction. One pivotal, hard-to-explain-away instance of this was the central role played by DNA in the "phage" experiments done by Hershey and Chase in the early 1950s.

Moving on to the King's College/Cavendish Laboratories phase of these developments (while setting aside most of the inter-human dynamic which speaks for itself), we find a familiar dichotomy. Technologically enhanced, minimal sensory data on the one hand, and interpretive thought models on the other. We have vague, subsensory X-ray diffraction data (with biochemical contributions) that is being interpreted by tinkertoy-reminiscent thought models. Things are so far away from the phenotypic sense world that, for much of this time period, the scientists were not even sure which kind of substance they were dealing with: protein or DNA?

One of the demands put on the thought models was that they must be able to explain how, during cell division, chromosomes are able to create exact copies (replicas) of themselves. For only then could they pass on their exact "content" to new cells (via mitosis) and to following generations (via meiosis). In 1953, finally, it appeared that the key to the kingdom had been found! Despite the extreme vagueness of the X-ray refraction mediated, subsensory side of the equation, Watson and Crick came up with a thought model (a tinkertoy-like double helix structure) that appeared to provide "the solution." From the structure of the double helix as they described it, it was apparent how exact replication would be possible. And from the sequence of bases it was now possible to imagine how the "primary" substance of the body (protein) could be synthesized. With this, the "central dogma" was born (DNA→RNA→Protein) and the "genetic code" revealed.

This was the "golden age of genetics," and soon thereafter, with enhanced technology, tons of money, and unbounded expectations, the HGP was born. Mapping the sequence of the three billion base pairs in the human genome was expected to create "the most important, most wondrous map ever produced by humankind," and to revolutionize medicine! After all, from the perspective of the "golden age" and HGP, the code of the human body was in the genome and the body's task was only to read and execute those instructions! So if this script of life could be enciphered, then we should be able to discover its weak spots, and with the help of genetic engineering probably change the script itself for the better! And as an added motivator, there would be lots of research dollars available and big profits to be made along the way!

As far as the kind of thinking involved in all of this is concerned, we see once again the triumph of reductionism—a way of thinking that breaks down complex phenomena into their constituent parts (in this case molecular parts) and then seeks to explain those phenomena out of the interaction of the parts.[106] Stephen Rothman speaks in this context of genetic determinism.

> *Genetic determinism is the belief that who and what we are is wholly determined by the genes, by the DNA, that we inherit from our parents. In this view, nothing less than life itself is the product of this inheritance.... From this perspective, the rest of biology—the rest of what we understand about life, human life not excepted—may be interesting, and worthy of understanding, but it is all commentary, mere footnotes to the main narrative about DNA and protein molecules.* (Rothman 2002, pp. 4–5)

106 Although not a topic that I would explore extensively with the students, it is important to bear in mind the role that funding and technology play in influencing the kind of thinking that gains prominence in a particular field of study. In his history of the molecular revolution, Michel Morange devotes an entire chapter to this topic. In it he emphasizes:

Most historians of science have argued that the Rockefeller Foundation had an essential role in the paradigm shift that took place in biology in the 1940s.... The grant-awarding policy followed by Warren Weaver and the Rockefeller Foundation, which prioritized funding for physics and chemistry laboratories that wanted to start studying biology, or biology laboratories that needed to buy equipment to make physical studies of organisms, was quite clearly based on a materialist, reductionist philosophy that saw future progress in biology as coming from an understanding of the molecules that constitute all life. Providing laboratories with the ability to use physico-chemical techniques undoubtedly played a key role in changing scientists' mentality.... Too often, a conceptual presentation of the history of science makes it impossible to see just how much the models and the concepts used by scientists are determined by the technologies employed.... In sum, the Rockefeller Foundation undoubtedly played an important role in the development of a molecular approach to biology that was based on the use of the techniques of chemistry and physics. This led, to the more or less unconscious validation of a reductionist conception of biology. (Morange 2020, pp. 83, 86)

In fact, we seem to have reverted once again into a way of understanding reminiscent of the preformation and mosaic theories we heard about early on in this block. In reference to this approach to biology, Richard Lewontin says:

> *It is really preformationism that has triumphed, for there is no essential difference, but only one of mechanical details, between the view that the organism is already formed in the fertilized egg and the view that the complete blueprint of the organism and all the information necessary to specify it is contained there.* (Lewontin 2000, p. 6)

Everything is seen as fixed from the very beginning (even if not already preformed as Epicurus envisioned it)—the code has already been written! Although now at a much higher level of sophistication, genetics seems to be revisiting the high point of chromosome mapping under T.H. Morgan, where the starting points of all traits "stand side by side, like stones of a mosaic."

But as we have already seen, the reductionist findings of the HGP ended up undermining the one-sided view of the human being (and of other living creatures) that was its driving force. The HGP helped us realize that focusing on structure and discrete elements is to overlook the dynamic, process-nature of organisms. Organic processes are so complex and interwoven that even apparently fixed and permanent structures are constantly being renewed and reworked to fit the situation at hand. Organism are verbs, not nouns! We saw this in the first part of this block when we considered findings in the field of experimental embryology. The study of heredity seemed to lead us away from that dynamic picture and into molecular structures that were to be the cause of everything else. (Although they needed the help of non-empirical metaphors such as information, text, instructions, code, etc. as descriptors.) But as it turned out, although these structural elements are involved in the organic processes, being involved is not the same thing as being the cause thereof. Stephan Rothman characterizes the reductionist attempts to make specific genes the causes of complex properties in whole organisms in this way:

> *What they have done is infer "sufficiency" while only having demonstrated "necessity." This is like believing that electricity produces the "content" of television programs, because the content disappears when we pull the plug.* (Rothman 2002, p. 97)

Rothman is pointing to the classic distinction between a necessary condition for something to take place and its cause. To roughly paraphrase Plato: When I get up to close the window, the movement of my legs is a necessary condition for me to be able to do that, but not the cause of my action. (The cause of my action can be found in my insight that the room is getting cold.)

I'll close this review by paraphrasing Ernst Michael Kranich with a final "musical" image that can help underline for the students the relationship between a "necessary condition" and a "sufficient cause": If a great pianist on her way to perform Beethoven's *Piano Sonata No. 8 in C minor, Op. 13*, falls off her bicycle and breaks the little finger on her left hand, the concert will have to be canceled. An extreme reductionist sitting in the audience as the announcement is made might immediately jump up and proclaim: "OMG, I have just grasped the cause of Beethoven's *Piano Sonata No. 8 in C minor, Op. 13*: It's the little finger on the left hand of the pianist—for without it the Sonata cannot appear, cannot be heard." (An even simpler variation on this motif could be: If a key breaks on the piano, the music cannot be played—does that make the key the cause of the music? Is the piano the cause of the music, or is it simply a necessary condition for it to be heard?)

In Conclusion

By taking the students through the unfolding of a science like genetics in such a way as we have attempted above, we are helping them realize that the "nothing-but-ness" judgments that we so easily carry around as snapshot takeaways from what we have learned or heard somewhere, do not do justice to the multileveled nature of reality. Real science is not a simple brick-upon-brick building up of knowledge, but is a many-faceted process that is strongly shaped by the dominant mindset (paradigm) that prevails in a particular field of inquiry at any particular time. In fact, awakening the students to the formative role that thinking plays in the prevailing picture of the world at any point in history, or in any particular culture, is a central takeaway from this block. That insight extends far beyond the realms of science, of course, and is essential—the students will realize—in understanding many aspects of their own lives.

Resources Parts I and II

Alberts, Johnson, Lewis, et al. (2007). *Molecular Biology of the Cell.* New York: Garland Science.

Baer, Hazen, Jameson, and Sloan (1971). *Central Concepts of Biology.* New York: Macmillan.

Beadle, G. and Tatum, E. (1941). "Genetic control of biochemical reactions." In *Neurospora. PNAS,* 27, 502. Retrieved from http://www.pnas.org/content/27/11/499.full.pdf

Berg, J., Tymoczko J., and Stryer L. (2002). *Biochemistry.* New York: W.H. Freeman.

Beurton, P., Falk, R., and Rheinsberger, H. (2008). *The Concept of the Gene in Development and Evolution.* Cambridge: Cambridge University Press.

Bird, A. (2018) "Thomas Kuhn." *The Stanford Encyclopedia of Philosophy.* Edward N. Zalta, ed.

Black, D. (1998). "Splicing in the inner ear: a familiar tune, but what are the instruments?" *Neuron* 20: 165–168.

Bonner, J. (1952). *Morphogenesis.* Princeton: Princeton University Press.

Buchner, Eduard – Biographical. NobelPrize.org. Nobel Media AB 2021. 16 March 2021. <https://www.nobelprize.org/prizes/chemistry/1907/buchner/biographic

Carlson, E. (2004). *Mendel's Legacy.* Cold Spring Harbor, NY: Cold Spring Harbor Laboratory Press.

Coleman, C. (2006). *Understanding Cancer.* Baltimore: Johns Hopkins University Press.

Cobb, M. (2015). *Life's Greatest Secret.* New York: Basic Books.

Cohen, I., Atlan, H., and Efroni, S. (2016). *Genetics as Explanation: Limits to the Human Genome Project.* Chichester: John Wiley & Sons, Ltd.

Crick, F. (1988). *What Mad Pursuit—A Personal View of Scientific Discovery.* New York: Basic Books.

Elmore, S. (2007). "Apoptosis: A review of programmed cell death." *Toxicol Pathol.*; 35(4): 495–516.

Fels, et al. (1976). *Der Organismus.* Stuttgart: Ernst Klett Verlag.

Flemming, W. (1880). "Beitrage zur Kenntnis der Zelle und Ihre Lebenserscheinungen." *Archive fuer Mikroscopische Anatomie*16: 302–406. Cited in Carlson (2004), p. 25.

Fox Keller, E. (2010). *The Mirage of a Space between Nature and Nurture*. Durham, NC: Duke University Press.

Francis, R. (2011). *Epigenetics*. New York: W.W. Norton.

Gilbert, S. (2010). *Developmental Biology*. Sunderland, MA: Sinauer.

Gould, S. (1977). *Ontogeny and Phylogeny*. Cambridge: Harvard University Press.

Grady, D. (2010). "A lasting gift to medicine that wasn't really a gift." *The New York Times* 9. April 10.

Gray, G. (1967). "The organizer." In Kennedy, D., ed. *From Cell to Organism – Readings from Scientific American*. San Francisco: Freeman & Co.

Grimm, D. (2020). "Is it time to replace one of the cornerstones of animal research?" *Science* AAAS (sciencemag.org), June 25, 2020.

Gruber, J. (2013). "The unfulfilled promise of genomics." In Krimsky and Gruber, eds. *Genetic Explanations – Sense and Nonsense*. Cambridge: Harvard University Press.

Gurtner, G., Werner, S., Barrandon, Y., et al. (2008). "Wound repair and regeneration." *Nature* 453, 314–321.

Hadorn, E. (1961). *Experimentelle Entwicklungsforschung*. Berlin: Springer-Verlag.

Harris, R. (1976). *Cancer*. London: Penguin Books.

Harold, F. (2003). *The Way of the Cell: Molecules, Organisms, and the Order of Life*. Oxford: Oxford University Press.

Hayflick, L. (1968). "Human cells and aging. *Scientific American*. Vol. 218, No. 3.

Hays, H. (1972). *Birds, Beasts, and Men*. Baltimore: Penguin.

Heusser, P. (2016). *Anthroposophy and Science*. Frankfurt: Peter Lang GmbH.

Hickman, Roberts and Larson. (2001). *Integrated Principles of Zoology*. Boston: McGraw-Hill.

Holdrege, C. (1996). *Genetics and the Manipulation of Life*. Hudson, NY: Lindesfarne Press.

Holdrege, C. and Talbott, S. (2008). *Beyond Biotechnology.* Lexington: University Press of Kentucky.

Hubbard, R. and Wald, E. (1993). *Exploding the Gene Myth.* Boston: Beacon Press.

Hubbard, R. (2013). "The Mismeasure of the Gene." Krimsky and Gruber, eds. *Genetic Explanations.* Cambridge: Harvard University Press.

Hurney, C., et al. (2015). "Normal table of embryonic development in the four-toed salamander, *Hemidactylium scutatum.*" *Mechanisms of Development.* Volume 136, 99–110.

Husemann, F. (1974). *Das Bild des Menschen als Grundlage der Heilkunst.* Vol. I. Stuttgart: Verlag Freies Geistesleben.

Illis, L. (2012). "Central nervous system regeneration does not occur." *Spinal Cord* 50, 259–263.

Jacob, F. (1998). *Of Flies, Mice, and Men.* Cambridge: Harvard University Press.

Kahneman, D. (2011). *Thinking, Fast and Slow.* New York: Farrar, Straus and Giroux.

Kaur, G. and Dufour, J. (2012). "Cell lines–valuable tools or useless artifacts." *Spermatogenesis.* 2(1): 1–5.

Keller, E. (2002). *The Century of the Gene.* Cambridge: Harvard University Press.

Kiani, A., et al. (2022). "Ethical considerations regarding animal experimentation." *J Prev Med Hyg* 2022; 63 (suppl. 3): E255–E266.

Kovall, R. and Blacklow S. (2010). "Notch signaling." In *Current Topics in Developmental Biology,* Vol. 92, 31–71.

Kraft, A.v. (1968). "Organismis, Zelle, Ultrastruktur." *Elemente der Naturwissenschaft,* 9, 45–55.

Krogh, D. (2010). *Biology.* Upper Saddle River, NJ: Prentice-Hall, Inc.

Kuhn, T. (1970). *The Structure of Scientific Revolutions.* Chicago: University of Chicago Press.

Lai, E. (2004). "Notch signaling: control of cell communication and cell fate." *Development* 131: 965–973.

Linder, H. (1998). *Biologie.* Braunschweig: Schroedel Verlag.

Lewontin, R. (1992). "The dream of the human genome. *The New York Review*. (May 28, 1992).

Lewontin, R. (2000). *The Triple Helix*. Cambridge: Harvard University Press.

Locke, D. (1969). *Enzymes—The Agents of Life*. New York: Crown Publishers.

Logan, T. (2014). "The Animal that re-grows its head." *BBC Future, Science & Environment*.

Magidson, V. and Khodjakov, A. (2013). *Methods in Cell Biology*. Volume 114, 545–560.

Maienschein, J. (2007). "To evo-devo through cells, embryos, and morphogenesis." In Laubichler and Maienschein, eds. *From Embryology to Evo-Devo*. Cambridge: MIT Press.

Marieb, E. and Hoehn, K. (2012). *Human Anatomy and Physiology*. San Francisco: Pearson.

Matile, P. (1973). "Die heute entscheidende Phase in der biologische Forschung." *Universitas*, Heft 5.

Michio, K. (1997). *Visions: How Science Will Revolutionize the 21st Century*. New York: Doubleday.

Moore, J. (2003). *Science as a Way of Knowing*. Cambridge: Harvard University Press.

Morange, M. (2020). *The Black Box of Biology—A History of the Molecular Revolution*. Cambridge: Harvard University Press.

_______. (2008). *Life Explained*. New Haven: Yale University Press.

_______. (2001). *The Misunderstood Gene*. Cambridge: Harvard University Press.

_______. (1998). *A History of Molecular Biology*. Cambridge: Harvard University Press.

Morgan, T. (1913). "The relations of genetics to physiology and medicine." In *Nobel Lectures…Physiology and Medicine*. 1922–1941, 313–328.

Mukherjee, S. (2010). *The Emperor of All Maladies – A Biography of Cancer*. New York: Scribner.

_______. (2016). *The Gene*. New York: Scribner.

Muller, H. (1950). "Our load of mutations." *The American Journal of Human Genetics* Vol. 2 No. 2.

NIH (2020). "Significant Research Advances Enabled by HeLa Cells." Office of Science Policy.

Nitschmann, J. (1986). *Entwicklung bei Mensch und Tier*. Braunschweig: Vieweg + Sohn.

MedlinePlus Genetics (2020). NOTCH1 gene. https//medlineplus.gov

Noble, D. (2008). *The Music of Life*. Oxford: Oxford University Press.

Okada, T. (1995). *Transdifferentiation –Flexibility in Cell Differentiation*. Oxford: Oxford University Press.

Pfeiffer, J. (1969). *The Cell*. Time, Inc.

Prabakaran S., Lippens G., et al. (2012). "Post-translational modification: nature's escape from genetic imprisonment and the basis for dynamic information encoding." *WIREs Syst Biol Med*.

Rappoport, J. (2017). *The Cell*. Dallas: BenBella Books.

Raven, Johnson, et al. (2016). *Biology*. New York: McGraw Hill.

Reddien, P. and Sánchez, A. (2004). "Fundamentals of planarian regeneration." *Annual Review of Cell and Developmental Biology*. 20: 725–757.

Roberts, L. (1992). "Why Watson quit as project head." *Science*, Vol. 256.

Romer, A. (1997). *The Vertebrate Body*. Philadelphia: W.B. Saunders.

Schmeck, H. and Schmeck, H. (1986). "HeLa's legacy." *The New York Times*, 15 June 1986.

Schotté, O. (1938). *Embryonic Development and Induction*. New Haven: Yale University Press.

________ (1940). "Xenoplastic induction of rana pipiens adhesive discs on balancer site of amblystoma punctatum." *Journal of Experimental Zoology*, Wiley Online Library (16 May 2005).

Schneider, L. (1964). "Morphogenese und Dynamik cytoplasmatischer Membranen." *Dtsch. ZooL Ges.*, 243–272.

Schnetter, M. (1968). Die Ära Spemann-Mangold am Zoologischen Institut der Universität Freiburg i. Br. *Ber. Naturf. Ges. Freiburg i. Br.* 58, 95–110.

Schwartz, J. (2008). *In Pursuit of the Gene.* Cambridge: Harvard University Press.

Serpente, N. (2011). "Cells from icons to symbols: molecularizing cell biology in the 1980s." *Studies in History and Philosophy of Biological and Biomedical Sciences*, 42: 403–411.

Shay, J. and Wright, W. (2000). "Hayflick, his limit, and cellular ageing." *Nat Rev Mol Cell Biol.* Oct;1(1): 72–76.

Shapiro, J. (2011). *Evolution—A View from the 21st Century.* Upper Saddle River, NJ: FT Press Science.

Shubin, N. (2009). *Your Inner Fish.* New York: Random House.

Skoot, R. (2010). *The Immortal Life of Henrietta Lacks.* New York: Crown Publishing.

Spemann, H. (1935) Nobel Lecture: "The Organizer-Effect in Embryonic Development," NobelPrize.org

Strathern, P. (1997). *Crick, Watson and DNA.* New York: Doubleday.

Sutton, W. (1902). "On the morphology of the chromosome group in Brachystola magna." *Biological Bulletin* 4.

_______. (1903). "The chromosomes in heredity." *Biological Bulletin*, 4: 231–251.

Tagarelli, Piro, Lagonia, and Tagarelli (2003). "Walter Stanborough Sutton: A hundred years after the chromosomal theory of heredity." *Chromosoma* 112: 1–5.

Talbott, S. (2010). "Context matters – the epigenetics revolution." In *Context* #23.

_______. (2013). "The myth of the machine organism: from genetic mechanisms to living beings. In Krimsky and Gruber, eds. *Genetic Explanations – Sense and Nonsense.* Cambridge: Harvard University Press.

Tsonis, P. and Fox, T. (2009). "Regeneration according to Spallanzani." *Developmental Dynamics*, Vol. 238/9.

Tsonis, W. and Del Rio-Tsonis, K. (1995). "Transdifferentiation as a basis for amphibian limb regeneration." *Semin Cell Bio.* June 6(3): 127–135.

Twitty, V. (1966). *Of Scientists and Salamanders.* San Francisco: Freeman.

Vellejos, C. (1994). In *DNA-based Markers in Plants.* Phillips and Vasil, eds. New York: Kluwer Academic Publishers.

Waddington, C. (1968). "The theory of evolution today. In *The Alpbach Symposium. Beyond Reductionism*. Koestler and Smythies, eds. Boston: Beacon Press.

Wake, M. (1992). *Hyman's Comparative Vertebrate Anatomy*. Chicago: University of Chicago Press.

Watson, J. (1968). *The Double Helix*. London: Penguin Books.

_______. (2017). *DNA–The Story of the Genetic Revolution*. New York: Alfred Noble.

Weiss, P. (1971). "The living system: determinism stratified." In *The Alpbach Symposium. Beyond Reductionism*. Koestler and Smythies, eds. Boston: Beacon Press.

Wickman, G., Julian, L., Mardilovich, K. et al. (2013). "Blebs produced by actin–myosin contraction during apoptosis release damage-associated molecular pattern proteins before secondary necrosis occurs." *Cell Death Differ* 20, 1293–1305.

Wilkins, M. (2003) *The Third Man of the Double Helix: The Autobiography of Maurice Wilkins*. Oxford University Press.

Williams, G. (2019). *Unravelling the Double Helix*. New York: Pegasus Books, Ltd.

Wirz, J. (2008). "Nicht Baukasten, sondern Netzwerk." *Elemente der Naturwissenschaft* 88.

Yannuzzi, D. (2004). *Gregor Mendel*. New York: Scholastic, Inc.

Part III.
ECOLOGY

Developing "Value-ability" in the Waldorf School from Kindergarten through High School

The Awakening of Ecological Awareness in 20th Century Biology

The next few pages introduce the emergence of an important concept within the field of ecology. It is not introduced here as potential content in the 11th grade ecology block, but as background material that describes a relationship to nature that is nurtured throughout Waldorf education. Although the expression "value-ability" is unfamiliar to most Waldorf teachers, what it refers to is alive in their teaching and points to a way of experiencing the world that provides a strong foundation for an ecology block such as the one that will be described here.

From its very beginnings, ecology—as the study of how organisms interact with one another and their larger environment—has been concerned with community, systems and wholes. The holistic/community-oriented perspective that permeated the work of early ecologists, such as Henry Cowles and Frederic Clements, provided fertile soil for what would later develop into the field known as environmental ethics.

Mid-20th century, Albert Schweitzer influenced popular understanding with his concept of "reverence for life." This reverence for life was not limited to human beings. For Schweitzer the powerful status that humans enjoyed in the natural community did not entail a right to exploit, but a responsibility to protect. Schweitzer's holism coincided remarkably with the concept of a biotic community developed by ecologists in that same period. The image of this gentle Doctor, who selflessly served the native people of what was then French Equatorial Africa, and who at the same time carefully stepped around an insect or helped a worm back into the soil, made a strong impression on people. In 1952 he received the Nobel Peace Prize; ten years later Rachel Carson dedicated her book *Silent Spring* to him.

One seminal thinker whose work played a decisive role in preparing the way for a change in environmental consciousness was Aldo Leopold. His enormous influence stems from a small manifest of 25 pages, "The Land Ethic," which appeared after his death in 1949 as the concluding essay of *The Sand County Almanac* (1949). Less than two decades later, it would become the intellectual touchstone for the birth of a far-reaching environmental movement in North America. In this essay Leopold addresses one of the central questions for the field of ethics, namely, how widely inclusive should our circle of moral consideration be? He does so already in the opening paragraphs of "The Land Ethic" with a very disconcerting image.

> *When God-like Odysseus returned from the wars in Troy, he hanged all on one rope a dozen slave girls of his household whom he suspected of misbehavior during his absence. This hanging involved no question of propriety. The girls were property. The disposal of property was then, as now, a matter of expediency, not of right and wrong. Concepts of right and wrong were not lacking from Odysseus' Greece: Witness the fidelity of his wife through the long years before at last his black-prowed galleys clove the wine-dark seas for home. The ethical structure of that day covered wives, but had not yet been extended to human chattels. During the three thousand years which have since elapsed, ethical criteria have been extended to many fields of conduct, with corresponding shrinkages in those judged by expediency only.* (p. 201)

According to Leopold, we as yet have no ethic that establishes our relationship as humans to land and the life it carries. Like Odysseus' slave girls, land is still property to us. Our relationship to it is solidly economic, one that entails privileges without obligations.

A land ethic, on the other hand, would extend the boundaries of the community to include animals and plants, waters and soils. It would transform the role of the human being from one of conqueror to member and citizen of the land-community. But how do we get from here to there, Leopold asked? For one, the content of conservation education must change. Rather than merely emphasizing adherence to the law and the practice of economic self-interest, we must teach a larger obligation to the health of the land. But obligations have no meaning without the development of conscience. It is herein, Leopold emphasizes, that the true task of education lies, for no significant change in ethics was ever achieved without internal change in our conviction and affection. A land ethic in Leopold's sense would reflect the existence of an ecological conscience that would, in turn, reflect a conviction of individual responsibility for a healthy land capable of self-renewal. Such conscience is born out of love, respect and admiration for land, out of the impulse to preserve the integrity, stability and beauty of the biotic community.

Whereas the appearance in 1962 of Rachel Carson's *Silent Spring*[107] is widely considered the spark that kindled the environmental movement in its current

107 If time allows, it is very interesting for the students to learn more about Rachel Carson and the impulse that led to her profoundly influential book *Silent Spring*. Jane Goodall's work with animals in Africa is also a very fruitful topic to pursue.

form, it was not until the early 1970s that serious philosophical reflection about ethical issues concerning the ways in which human beings should interact with the nonhuman natural world began.[108]

After reading Leopold's "The Land Ethic," many environmentalists grew increasingly dissatisfied with instrumental arguments that had been used to justify the protection of nature based on human use and benefit.[109] This led to a fundamental distinction in environmental ethics between instrumental and non-instrumental value. "Instrumental value" is assigned to something based on its usefulness—it serves as a means to another end. Water has instrumental value because all humans would die without it; a tree has value because it supplies lumber for furniture; a worm is valuable because it enriches the soil. Our own existence, on the other hand, is not viewed as a means to some other end, it represents an end in itself; human life possesses "non-instrumental" or "intrinsic" value.

Instrumental value can and has been interpreted very broadly. The natural world can be viewed as providing humans with a range of spiritual, aesthetic and physical values, for example. An instrumental perspective would speak against the reckless exploitation of the environment in light of its value for future human generations. Most discussion of environmental issues at the international level—ozone depletion, global warming, rainforest destruction, etc.—focuses ethically on the effects they have on human beings. Concerns about resource management make up virtually the entire ethical agenda at the level of popular political debate, and the discussion of sustainable development is frequently limited to resource management.

Such instrumental approaches to environmental value have been the object of frequent criticism from within the field of environmental ethics. One of the primary aims of much work within that field has been, in fact, to contest the instrumental valuation of nature and to argue that new ways of considering and valuing it are necessary. In the 1970s and early 1980s, the first round of anti-instrumental critique took the form of suggesting that nature might have—or should have—rights.

108 The first conference on environmental philosophy was held in 1971 at the University of Georgia. The possibility that natural things and areas might be considered objects of moral concern launched a variety of schools of environmental ethics.

109 One of the most important distinctions in U.S. environmental history is between the environmental movement as it emerged in the 1960s and the conservation efforts that preceded it. Conservation did not refer to protecting or preserving nature; it meant the wise and efficient use of natural resources in service of the long-term material interests of humankind. Half a century after its inauguration, conservation philosophy came under heavy attack from a new breed of environmentalist who saw much of the work of the Forest Service and the Bureau of Reclamation as infringing upon the rights of nature itself (Nash 1989).

Holmes Rolston, one of the leading thinkers in regard to questions concerning the intrinsic value of nature, calls the human capacity to take real interest in something—to value it in and of itself—"value-ability." When we value a great sequoia tree, for instance, we value it intrinsically if our valuing relates to the tree itself, intrinsically, without reference to its worth as a provider of wood or shade. We can develop the ability to value the welfare of endangered species and the preservation of natural habitats for their own sake, and that it is, in his view, the task of environmental education to help individuals develop such capacity.

Rolston describes ecosystems as generators of an order that envelop and bring forth the beauty, richness, integrity, and dynamic stability of their component parts. An ecological system represents a field with characteristics that are as necessary for life as any properties found in an individual organism. The matrix of the earth's ecosystems is the womb in which life on earth evolved and thus is also—in addition to the individual organisms contained within it—an appropriate unit for moral concern, is an entity that contains value in and of itself.

In an influential article, "The Power and Promise of Ecological Feminism," Karen Warren argues that we should conceive ourselves to be "in relationship with" others, including the non-human environment. She demonstrates this with a first-person narrative about rock climbing. After an opening section that portrays a day of struggle trying to *overcome* the rock in an attempt to reach the top and complete the climb, she contrasts this with a very different experience on the second day.

> *I heard a cacophony of voices—birds, trickles of water on the rock before me, waves lapping against the rocks below. I closed my eyes and began to feel the rock with my hands—the cracks and crannies, the raised lichen and mosses, the almost imperceptible nubs that might provide a resting place for my fingers and toes when I began to climb. At that moment I was bathed in serenity. I began to talk to the rock in an almost inaudible, childlike way, as if the rock were my friend. I felt an overwhelming sense of gratitude for what it offered me—a chance to know myself and the rock differently, to appreciate unforeseen miracles like the tiny flowers growing in the even tinier cracks in the rock's surface, and to come to know a sense of being in relationship with the natural environment. It felt as if the rock and I were silent conversational partners in a long-standing friendship. I realized then that I had come to care about this cliff which was so different from me, so unmovable and invincible, independent and seemingly*

> *indifferent to my presence. I wanted to be with the rock as I climbed. Gone was the determination to conquer the rock, to forcefully impose my will on it; I wanted simply to work respectfully with the rock as I climbed. And as I climbed, that is what I felt. I felt myself caring for this rock and feeling thankful that climbing provided the opportunity for me to know it and myself in this new way.* (Warren 1990, p.131)

Warren emphasizes that this narrative gives voice to a "felt sensitivity," that it speaks of being "in relationship with" the non-human environment that is often lacking in traditional analytical ethical discourse. Such a narrative also shows the difference in attitude between an imposed, conqueror-type relationship to a rock, which is all about making it to the top, and an emergent, caring relationship that grows out of felt, lived experience.

Warren does not exclude considerations of rights, rules or utility—for there are situations where these are appropriate or useful—but provides a central place for values such as care, love, respect, appropriate reciprocity, and wonder, that typically go unnoticed or are underplayed in traditional ethics. By making a central place for such values, ecofeminism presupposes that our relationships to others are central to our understanding of "who we are." It thus gives voice to the sensitivity that in climbing a mountain we are doing something in relationship with an "other," an "other" whom we can care about and treat with respect. Ecofeminism understands relationships to be something intrinsic to who we are, not an add on to human nature.

One leading ecofeminist, Carolyn Merchant, sees the abstract, distanced, "objective" relationship to nature that characterizes modern western consciousness arising with the founders of modern science. She links it to the birth of the mechanistic view of nature developed in the 17th century, a view that assumes nature can be divided into parts and that these parts can be rearranged to whatever (instrumental) end we might find useful. In his highly influential book, *Earth in Balance*, Al Gore also argues that the outer environmental crisis is, in fact, a reflection of an inner crisis of human consciousness, an inner crisis that has its origins in the dualistic, reductionistic tendencies of modern science. Gore speaks of a pivotal shift in our relationship to nature that began with thinkers such as Bacon, Galileo and Descartes. A consequence of their approach to the sense world was the exclusion of all qualitative human experience (color, sound, etc.) from scientific deliberation. Direct human sense experience was shunted aside as merely subjective because it was not quantifiable. Another central aspect of the inner (non-spatial) human being —the life of feelings—was also devalued as purely subjective and thus of no significance in reference to the outer world. Gore contends that the healthy human being longs for a form of communion with the immediacy of living nature that involves the richness of sense experience and the engagement of our feelings. With our feelings excluded and our direct sense experience declared invalid, he says, the modern human being suffers a deep sense of loss and emptiness.[110] (Alexandra 1996; Gore 1992; Hargrove 2003; Katz 2003; Leopold 1949; Light and Rolston 2003; Merchant 1989; Nash 1989; Palmer 2003; Warren 1990; Warren and Cheney 2003)

110 Gore sees this emptiness as masked to a large degree by the addictive consumerism that represents one of the driving forces behind the extreme depletion of the earth's resources.

Gardening in Early Childhood and Grade School

Laying the foundation for "value-ability" in kindergarten and grade school

In contrast to the devaluation of sense experience and our feeling-life that Gore speaks of, Waldorf education offers a great deal that supports the development of "value-ability" and "being in relationship with the natural environment" from early childhood on. A garden is part of most Waldorf schools, it is a place where the children learn about caring for the soil, sowing seeds, planting, watering, harvesting, and much more.[111] Through many activities they learn to accompany the unfolding of nature through the seasons, harmonizing their own activities with the rhythms of nature: You don't simply sow seeds when it is convenient, but when "the time is right."

Preschool children live in a world of awe and wonder about what they meet in nature. For them everything is alive. They do not experience a separateness from the natural world. They are full of questions that their teachers respond to with imaginative stories and verses that highlight what they experience. Their thinking is not yet awake, but their feeling life resonates strongly with the environment around them and is nourished by an attitude of respect and reverence for that world.

They walk, for example, about the school garden on a regular basis. They see the constant changes taking place there and do—following clear guidelines from the teacher, of course—a bit of tasting, now and then. Larger fruits can be cut up by the teacher and handed around for them to try. On some occasions they might pick mint leaves and with the teacher's help make a pot of tea (with honey) that is placed in the sun to brew. After a walk about, chores can follow: feeding the worms and chickens, watering the plants with gardening cans. In the fall apples can be gathered and brought to the classroom, where the children help the teacher cut and dry them for storing in jars as the daily snack, or for cooking apple sauce on the stove. And lastly, in the late fall, the garden is put to bed with blankets of straw that the children place upon mother earth. It is her time of rest as the cold of winter approaches. "Bulb children" (daffodils, narcissus, etc.) can be planted—and since

111 An outstanding resource to this topic is *Growing Sustainable Children – A Garden Teacher's Guide*. (2017), by Ronni Sands and Willow Summer. The examples given above stem largely from that book, and from observations of and consultation with the gardening teacher at the Chicago Waldorf School, Patricia Favre Holdrege. For German speakers, Christoph Kaiser's *Gärten der Zukunft* (2013), Verlag Freies Geistesleben, is also a wonderful resource.

Fig. 3.1 a. Compost making; b. Planting; c. Watching things grow. (Photos: Barb Ciurej)

they are among the first to awaken in the spring, the children look forward to that moment with great anticipation.

Year-round the children gather food scraps after classroom cooking and feed them to the compost pile or to the worms in their worm bins. They care for the compost, making sure it receives enough water and a balanced diet, and see to it that the worms are properly covered and watered in addition to receiving all the food they need. Once everything has been transformed back into soil, the children bring it to the garden, where it once more becomes part of the cycle of life that mother nature displays to them through the course of the year.

Moving into the elementary school, the engagement in the garden continues in a similar manner, but gradually takes on new forms as the children move from learning primarily through imitation into a new kind of relationship with their teachers. By the time they reach third grade the children are beginning to experience themselves as independent beings, more self-aware than hitherto, but also of the roles that others play in their lives. They begin to realize that so much of what they depend on—housing, food, clothing, etc.—comes from the work of others. This makes it an appropriate time for them to learn about and even participate in such processes themselves. At its foundation, the third grade curriculum is filled with activities such as cooking, baking, house building, fiber processing, farming, and so on. In addition, the third graders usually take a week-long trip to a working organic (usually biodynamic) farm, where they experience firsthand the demands of farm life: working with the animals, harvesting crops, doing field work, etc.—an unforgettable experience for children at this age.

Because gardening is so central to the 3rd grade curriculum, songs, stories, and verses learned by the class often accompany activities in the garden. Farm songs have always been an integral part of agrarian life and can be sung during various work projects. Not surprisingly, gardening in the 3rd grade synchronizes with the seasons. For example, starting with the fall and going through the whole school year, the class follows the cycle of wheat from seed to bread. Last year's 3rd grade has already grown the wheat that this years' class is able to thresh and winnow in September. The wheat is then ground into flower, made into bread and baked in an oven. The class also sows the wheat for next year's 3rd grade. In June, when the wheat is tall and ripe and the seed-containing heads have formed, the children cut and bundle them into sheaves that they hang up to dry for summer.

Fig. 3.2 Harvesting. (Photos: Barb Ciurej)

Onions, potatoes, radishes and corn can also be harvested in the fall, each of which provides a different experience. Ferreting out a large, beautiful potato from below the surface is a very different experience from pulling down corn ears and removing the husks. As a staple food of Native Americans and early settlers, there are many legends, stories, songs and other activities involving corn that can be taken up as the year moves toward Thanksgiving. Many teachers also research indigenous practices in their area and develop lessons around these. Before winter comes, the children sow a cover crop, which is usually accompanied by a fitting verse.

Winter activities include processing milk—ideally from a local farm—into butter, cheese, and yogurt; carding wool and spinning it with a drop spindle, and building bird houses. In the spring the children will plant potatoes, corn, pumpkins, cucumbers, and much more. Making scarecrows is also an activity that the children enjoy greatly.

Without going further into the elementary school curriculum, it should be evident from the foregoing that in the early grades the children's natural tendency to develop a relationship with nature based on "care, love and respect" (Warren) can be nurtured and developed. The children carry a sense of connectedness and a feeling for what it means to work in harmony with nature, with the seasonal contractions and expansions that are interwoven with the path of the sun. This provides a healthy and feeling-permeated foundation for the next phase of life when their intellect awakens and they begin to view things from a greater distance, seeing clearly that they live in a world that is facing very serious and tenacious environmental issues.

If we were to confront them with such issues at an early age, it would overwhelm them and undermine their sense of confidence and trust in the world that they are growing up into. As they move toward and into the high school years, on the other hand, they want to know about, to understand and confront the issues that will face humanity in their lifetimes. And it is here that the high school has one of its central tasks: to help the students look such problems straight in the eye, but to do so in a way that does not turn them into cynics. Instead, they need assistance in learning to see pathways into the future that they can say yes to and support.

Taking Things Further

The 9th Grade Agriculture Block and Practicum

Many Waldorf high schools have a 9th grade agriculture block combined with a two-week farming practicum.[112] Depending on scheduling possibilities, the block itself (main lessons) can be crafted in different ways, falling before the practicum, after it, or perhaps allocating some time both before and some more after the farm experience. Depending on such timing issues, the way the material is framed can vary considerably.

Once the 9th graders have arrived at the farm—which they may already know from earlier grade school visits—it is important to awaken them to the larger picture of what they will be exploring in the days ahead. This can begin in an open area where the path of the sun can be visualized, not only as it moves across the sky at present, but also over the course of the year—to picture how its arc gradually moves northward from its low point at the winter solstice, until reaching its zenith in late June. Other large-scale factors that will play a role in the life of the farm can also be considered, such as the proximity to mountains or open plains, and how these relate to the prevailing winds and otherwise affect the weather at this latitude. Is there a coastline nearby with warm or cold currents, or are we sitting far out in the middle of a large continent? All of the above will have an influence on the amount and seasonality of precipitation in the area, as well as temperature ranges throughout the year.

Moving from such large-scale factors to more regional ones, questions of general soil characteristics can be considered—a familiar topic from their geology block earlier in the year. Sites in the midwestern United States, for example, benefit from rich soil layers that were transported from regions farther north during times of significant glacial activity. Nonetheless, local soil properties can vary considerably within a small area. A farm will have habitat variations that the farmer needs to understand and take into consideration when determining what will take place where. Steep areas with poor soil often lend themselves to forestry, steep wet areas with uneven surfaces to grazing, fertile lowland soils in mild climates to vegetable growing, etc. After reflecting on such relationships, many questions often arise for the students. What is the importance of streams and ponds for the life of the farm,

112 For schools in cities, this is often includes visits to various initiatives engaged in urban agriculture.

for example, or why are certain meadows used for hay, and what determines the location of grain fields, etc? And beyond such questions, we can inquire whether things always stay the same, or are there planned changes from year to year—and if so, why? Ideally a long walk with the farmer through various habitats can provide answers to questions of this kind and provide insight into how the current constellation of farm activities is habitat-based, both in the moment and over time.

It is also important to reflect on the interactions between the plants and animals living on the farm and how they relate to the surrounding landscape. Bushes, trees, the herbaceous layer, mammals, insects, birds, etc. form a community. It makes a huge difference for that community if the landscape is interspersed with sufficient grassland and woods. It makes a difference if pastures are surrounded by bushes and shrubs, if the bank of a stream is stabilized by the roots of trees, if a pond is surrounded by a belt of reeds and if a marsh-lowland exists in the area. Such elements support a lively and diverse ecosystem that a farm can contribute to and flourish within. It is also of great value to pay attention to the so-called edge effect. At the boundaries where habitats connect a greater diversity of organisms can be found than elsewhere.

From all their earlier gardening experiences, the students know, of course, that soil characteristics and quality can vary greatly. Connecting to their geology block, the teacher can discuss with the students how the silica, clay, and lime contents of the soil—depending on their prevalence—support different kinds of processes or hinder them.[113]

This leads us, then, to the diversity of organisms within the soil: to the decomposers, that break down organic material; the diggers, that keep the soil in circulation by mixing lower mineral layers with upper organic ones; and the bacteria, that liberate the essential elements from the soil particles, as well as the various predators (protozoa, mites, centipedes, beetles, birds, etc.) that keep the population of decomposers and diggers in check. In a healthy soil, all these organisms work in harmony with each other to maintain the balance between processes of growth and decay. Among the most abundant soil animals are the arthropods. Well over a billion may live on an acre of pasture soil, and even more in hardwood forests, where pathway-permeated layers of litter and soil provide them with excellent habitats.

113 Quartz, for example, promotes aeration and warming of the soil, clay promotes the supply of nutrients and water to plants, and lime, together with other alkaline elements, stabilizes humus substance and soil structure (Koepf 1990).

The number of organisms that live in the soil is mindboggling and hard for the students to imagine.

One Danish study found that a single cubic meter of soil contained thousands of small earthworms, 50,000 insects and mites, and 12 million round worms. In a single gram of soil, they found 30,000 protozoa, 50,000 algae, 400,000 fungi, and billions of individual bacteria. It is this amazing biodiversity that rejuvenates and maintains soil fertility. It is estimated that the live weight of soil-dwelling organisms (edaphon) varies between 4,500 and 22,000 pounds per acre, which is the equivalent of 3 to 16 cows per acre! In other words, a vast portion of the "livestock" that lives on a farm is out of sight (but hopefully not out of mind)—hidden in the soil while working tirelessly at its transformation. The yearly production of earthworm castings (dung) in healthy soil amounts to between 25 and 36 tons per acre! Some sources even say that earthworms produce one ton of castings per day per acre during the growing season! Such castings contain in excess everything that plants need for healthy development—nitrates, phosphates, potassium, etc.—and in a condition which allows for optimum uptake and processing (Koepf 1990; Nardi 2007; Raven et al. 2004; Shiva 2016; von Heyer 1976).[114]

As Herwig Pommeresche, author of *Humusphere: Humus, a Substance or a Living System?* (2020), puts it when speaking to farmers:

> *These numbers reveal how much "livestock" you have hidden on your farm, without ever seeing or noticing it. If you converted this quantity into actual livestock, how much would you have to supply them with on your farm in terms of feed, stalling, ventilation, weather protection, and eventual excrement disposal? How about veterinary costs, consultation, and researching?*

This web of life, hidden from us but ever present in the soil, does not end here. Although the students are well aware that plants are rooted in the earth, they generally have no idea of the extent and significance of this subsurface dimension. Once plants get established in the soil—with the help of all the edaphon living there—they contribute to the process of soil formation by breaking down solid rocks. Their roots are not only strong enough to crack stone layers, but they also release potent

114 Fresh earthworm castings are 5x richer in soluble nitrogen, 7x richer in soluble phosphorous, 11x richer in soluble potash, 2x richer in magnesium, and many times richer in trace elements than artificially fertilized soil (Raven et al. 2004; Shiva 2016; von Heyer 1976).

compounds that gradually transform some of the largest and hardest rocks into fine mineral particles.[115] After they die, roots not only leave behind organic matter in the soil, but as they decompose, they leave behind a legacy of vacant passageways.

In its early days, most of a plant's energy goes into growing downward. Small roots branch off of larger ones, and tiny root hairs branch off from them. Even in plants that are well established above ground, the roots usually outgrow and outweigh all the stems, leaves, flowers and fruits that are produced aboveground. A typical midwestern prairie produces three tons of roots for every ton of shoots. As they wend their way among soil particles and along the tunnels provided by earthworms, roots can grow to phenomenal lengths. After only four months, one rye plant can extend fifteen million roots, whose total length reach approximately 380 miles. From the perspective of surface area, these roots come in contact with 2,554 square feet of soil. (If the lengths and surface areas of the miniscule but countless root hairs covering every root are taken into consideration, the total length grows to roughly 7,000 miles and the surface area to 7,000 square feet.) As they navigate their way, roots insert themselves into passages between particles of soil. And it is no small power that they have at their disposal: They can exert a force of up to 150 pounds per square inch, which enables a root to crack and even lift a slab of concrete weighing hundreds of pounds. Or, as Nardi (2007) points out: "A root measuring four inches in diameter and three feet in length can lift 50 tons."

Roots of different plants not only occupy different depths in the earth, but they can also be very choosy about the soils they associate with, growing well only where they find a fitting assortment of nutrients. Not only that, but often those nutrients can only be accessed if suitable mycorrhizal fungi[116] are present to help them tap into those resources. In fact, almost 80 percent of all plants team up with these fungi in special partnerships that benefit both plant and fungi. In these instances, long strands of fungi (hyphae) surround the plant's roots and then extend well beyond the reach of the roots. The fungal strands can grow into tinier soil spaces than the roots and tap resources that would otherwise be inaccessible to the plant. Not only that, but the total length of fungal strands in one cubic meter of soil can reach 12,500 miles, which allows the plant to absorb from a much larger volume

115 For example, roots release CO_2 by breaking down the sugars they receive from the green plant above. The carbon dioxide reacts with water in the surrounding soil, forming carbonic acid. The latter reacts with rocks like granite, releasing nutrients and forming clay and silt.

116 From Greek: *myco* = fungus, *rhiza* = root.

of soil than its roots alone could do. These fungal strands are particularly good at recovering water and essential elements, in particular phosphorous, from the soil.

On the other side of the equation, the plant shares the sugars it produces above earth with the fungi below. This enables, among other things, the fungi to bring together thousands of filaments to form the varied mushrooms that appear above ground. The thousands of spores that each mushroom produces then travel by wind and animals to new sites where they settle in and form new partnerships. (Graham et al. 2006; Nardi 2007; Raven et al. 2004)

Composting

After considering this web of diversity, it is interesting for them to return to an aspect of gardening/ agriculture that they have been involved with throughout their Waldorf Education (primarily at the feeling and will level): the compost pile. Now, in the context of what they have learned about the life of the soil, they can meet more consciously and understand better their old garden companion. So, we ask them—based on what they have just learned—what do they think the compost pile is all about? After some discussion, complemented by "teacher knowledge" when needed, a rich picture arises of how compost represents an enhanced microcosm of what takes place through the activities of decomposers, diggers, bacteria and predators in the soil of forests, fields and prairies every day: the conversion of plant matter to humus. Letting the students stir up a compost pile can reveal a great number and variety of decomposers, diggers and predators hard at work. The compost pile provides a refuge for these creatures and recycles much organic material, giving back to the soil what would otherwise end up in a landfill. They can now better understand how the amazing heat levels (up to 165 degrees F) that compost piles can reach is generated by populations of bacteria, actinomycetes, and fungi that can reach densities of 10 billion microbes for every gram of compost.

Many important aspects of effective composting can also be discussed with the students, such as finding a good location and deciding on the optimal form and size of the pile. They learn that a pile that is too small will lose a lot of heat, while one that is too large can become too hot or receive too little air at its center. Only a well-aired compost undergoes thorough decomposition, so it is important to have

Fig. 3.3 Steaming compost pile.
(Andrew Dunn. CC BY-SA 2.0)

a coarse bottom layer (corn stalks, for example) that allows air to enter from below. Turning the pile weekly after it has reached its highest temperature also helps air circulation and hastens decomposition through enhanced oxidation—which also keeps the pile from stinking![117] Also not to forget is that decomposers do their best work when the compost is uniformly moist, but not water logged. An ideal water content would be similar to that of a squeezed sponge.

Echoing back to their organic chemistry block earlier in the year, a discussion can also follow about the materials that are put into the compost pile. The greater the variety, the greater the diversity of microhabitats provided for the decomposers. In particular, it is important to remember that the decomposers need a certain proportion of carbon to nitrogen so that they can multiply and continue to do their work. In general, materials that are green and moist tend to be high in nitrogen, whereas those that are dry and brown are high in carbon. Ideally, compost has a carbon to nitrogen (C:N) ratio between 10:1 and 30:1. Since nitrogen tends to be in short supply, it is important that nitrogen-rich manure and green vegetation such as grass clippings, plant cuttings, fruit and vegetable scraps are added to the compost. Typical C:N ratios for various materials are as follows.

Materials High in Carbon	**C:N**
autumn leaves	30–80:1
straw	40–100:1
wood chips or sawdust	100–500:1
bark	100–130:1
mixed paper	150–200:1
newspaper or corrugated cardboard	560:1

Materials High in Nitrogen	**C:N**
vegetable scraps	15–20:1
coffee grounds	20:1
grass clippings	15–25:1
manure	5–25:1

Source: Dickson, N., T. Richard, and R. Kozlowski. 1991. *Composting to Reduce the Waste Stream.* Northeast Regional Agricultural Engineering Service

117 When there is a lack of oxygen, anaerobic microbes take over the decomposition, releasing scent-exuding compounds like butyric acid (stinky-feet smell!), hydrogen sulfide (rotten-egg smell!), and ammonia.

Since decomposers thrive best with roughly equal portions of carbon-rich and nitrogen-rich materials, experienced composters add them to the pile in alternating layers that are about four inches thick. To facilitate this process, the two groups are collected and stockpiled separately before adding them to the pile simultaneously in layers.

How do we know when the compost is ready? It should resemble dark crumbly topsoil and have a pleasant earthy smell. The final exam can take place as follows: 1) Place some of the compost in a sealed plastic bag. 2) Wait a few days. 3) Smell it. If it smells rotten, it's not ready—put it back! If there is no smell, it's done— celebrate! (Cornell Compost; Nardi 2007; NYC 2009; Raven et al. 2004)[118]

118 If biodynamic preparations are used in the compost, the teacher can characterize the preparation plants and their habitats, assisted by the book: Extraordinary Plant Qualities for Biodynamics by Jochen Bockemuehl and Kari Jaervinen (2006). Other valuable resources in this regard are available from the Josephine Porter Institute for Applied Biodynamics (jpibiodynamics.org) and from the Biodynamic Association (www.biodynamics.com).

Factory Farming

As has often been mentioned, one central and very effective way of delving into a topic is use comparisons. Sometimes this is done in a subtle way without emphasis, sometimes it is done very explicitly. After all that has been described above, one possible way to conclude the study of agriculture in the 9th grade is by comparing very explicitly the ecologically-oriented approach to farming that the students have come to know over many years—and which has now been deepened by the 9th grade block/practicum—with what is often referred to as factory farming or industrialized agriculture. There are, of course, multiple levels that can be considered in such a comparison. Among them are the effect that each approach has on the larger environment (like the creation of a massive "dead zone" in the Gulf of Mexico), the quality and diversity of the food produced, and the way that animals are treated.[119]

It is shocking for the students to realize that on today's factory farms animals are crammed by the thousands into filthy, windowless sheds and cages, where they are often unable to turn around or lie down comfortably. Antibiotics are then used to make them grow faster and help keep them alive under the unsanitary conditions they live in. Many don't even experience the warmth of the sun or breathe fresh

Fig. 3.4 Wheat harvest in Northern Idaho.
(Agriculture in the United States. Wikipedia)

119 Helpful resources that explore various aspects and consequences of factory farming are: *Fatal Harvest—The Tragedy of Industrial Agriculture* (2002), edited by Andrew Kimbrell, *Tomatoland* (2011) by Barry Estabrook, and *Animal Factory* (2010) by David Kirby. Vandana Shiva's *Who Really Feeds the World?* (2016) can also be used in regard to the failures of agribusiness and the world food crisis.

air until the day comes when they are crammed into trucks to be transported to slaughter houses. I won't go further into this topic here, but simply let a few images point to what might come up for discussion through such comparisons.

Fig. 3.5 Irrigation of wheat in Yuma County, Arizona.
(Source: https://photogallery.nrcs.usda.gov/Index.asp. Public domain)

Fig. 3.6 Pregnant pigs and their piglets are forced to live in unrelenting confinement and misery.
(Source: Mercy for Animals Canada. CC BY-SA 2.0 Generic)

Fig. 3.7 Hens in a battery farm.
(Maqi. CC BY-SA 3.0 Unported)

Harvesting from Earlier Life Science Blocks

After focusing on phenomena explainable through straightforward linear causality in the 9th grade human biology block, in the 10th grade we challenged their growing thinking-capacity in a new way as we moved into complex phenomena—such as those found in the cardiovascular system—that required a much more dynamic understanding of living organisms. This culminated at the conclusion of the block when various organs and processes were compared with each other. With the help of such comparisons, it was possible to develop the concept of "polarity," which sheds light on a fundamental organizing principle within organisms.[120]

It became clear that a polarity is composed of two opposing or contrasting tendencies that work together, such as in the polarity between the digestive tract as it breaks down substances, and the liver as it builds them back up again. The poles need each other in order to exist, in order to do what they do and to play the role they play within the larger whole in which they are embedded. The students have also learned to see how polarities are not only based on contrasting organs and organ systems, but also have a temporal component. They have not only seen this within individual organs—the systole and diastole of heart, for example—but also through the countless rhythms that inform the natural world and our daily lives: summer and winter, day and night, sleeping and waking, effort and rest, and so on.

Through such an overriding principle that manifests in countless ways, it is possible to bring life not only into the student's understanding of individual organisms, but also into the way they comprehend nature as a whole. For this reason, such considerations are an excellent preparation for seeing ecosystems with new eyes, for realizing that living organisms and their environments are much more verbs than they are nouns. They are lawfully organized processes that manifest in time through the rhythmical balancing/harmonizing of a multitude of functions.

The "developmental biology portion" of the 11th grade biology block discussed earlier also provides valuable perspectives for understanding ecosystems. In that block we learned that organisms are not preformed, but follow a lawful process of unfolding in time through a regulated process of cell division and differentiation—even cell death (apoptosis). Within all of this there is considerable flexibility if conditions call for it. At the cytological level we also found that the

120 See Holdrege (2022).

cell's ultra-structures are subordinate to the activity (needs) of the cell at any given moment. We even saw how organisms are capable of returning to an earlier phase of development when injured and to regenerate a part of the organism so that it corresponds again to the larger whole (morphogenesis). During this process cells not only dedifferentiate and re-differentiate, but some even trans-differentiate. It also became clear that what we know as cancer is a manifestation of cells that no longer subordinate themselves to the larger whole.

The study of genetics seemed at first to take us away from the organism as a whole and to "disappear" into subsense-perceptible realms that had to be interpreted with the help of complex and frequently quite hypothetical thought models. However, as we turned into the 21st century it became more and more evident that, in the words of Steve Talbott:

> *The fixity, precision, and unambiguous logical relations of the code... has lately been dissolving more and more into the fluid, dynamic exchange of living processes. Organisms, it appears, must be understood and explained at least in part from above downward, from context to sub-context, from the general laws or character of their being to the never fully independent details.* (Talbott 2010)

With this as a background, we move in the second 11th grade block to the other end of the life science spectrum by shifting from the microscopic to the macroscopic, where organisms are considered in their interactions with each other and within the larger biosphere in which they are all embedded.[121]

121 The 9th grade geology block, the 9th grade chemistry block, and the 10th grade geophysiology (meteorology) block—all of which have been mentioned previously—also contribute importantly to the enriching background that prepares for the block under discussion.

11th Grade Ecology Block

There are, of course, countless entry points and areas of possible emphasis when studying ecosystems and the web of relationships that they reveal. In the limited time available during a three- or four-week main lesson block, however, choices must be made. In the spirit of Wagenschein's "exemplary method," I have found it fruitful to focus on a few areas and to seek in them characteristic phenomena that build upon what the students have learned in previous blocks, while revealing fundamental principles at work in ecosystems around the earth.

The Intertropical Convergence Zone—A Central Component of the Biosphere[122]

One effective way to begin the study of ecosystems is to *start big*. In their 10th grade earth science block that considers the earth as a whole, the students usually learn about the Intertropical Convergence Zone (ITCZ) and Hadley Cells.[123] If so, a quick review can refresh their memories and provide the basis for looking at the three dominant biomes found on the continent of Africa. If they have not yet learned about the ITCZ, it can be introduced in a manner similar to the following.

We remind the students that somewhere near the theoretical line known as the equator, there appears—when averaged over time on satellite maps—a white girdle of clouds, a shining ring around the planet that is set against the dark oceans below.

What is creating this cloud girdle? The students will remember how trying to comprehend this amazing phenomenon in the 10th grade, they were led back to the path of the sun. When overhead, it warms the earth more intensely than when it shines through the atmosphere at a more oblique angle. This phenomenon explains why areas of the earth near the equator (the tropics) are warmer than elsewhere. When heating, the hot tropical air near the earth's surface rises with vigor, thereby drawing in air to replace it. In this way, winds are pulled from the northern and southern hemispheres toward the equator for several thousand miles. Laden with

122 *Biosphere* is defined by the *Oxford Dictionary of Ecology* (1998) as "the part of the Earth's environment in which living organisms are found, and with which they interact to produce a steady-state system, effectively a whole-planet ecosystem. Sometimes it is termed *ecosphere* to emphasize the interconnection of the living and non-living components." The study of the biosphere is often referred to as *geophysiology*. It is defined as "the scientific study of the earth as a total system that comprises all its biotic and abiotic components (i.e., the biota, atmosphere, hydrosphere and lithosphere)" (Allanby 1998).

123 This block is often titled "Meteorology" or "Geophysiology."

Fig. 3.8 Band of clouds created by the ITCZ that circle the earth near the equator. (Source: http://earthobservatory.nasa.gov/IOTD/view.php?id=703)

vapor swept up from the sea, these winds converge at the equator, which causes them to switch direction from the horizontal to the vertical. Rapidly ascending, this hot tropical air cools as it rises. The moisture it carries begins to condense and clouds arise in an earth-girdling ring of white. As it moves slowly along its westward path, this ring of moisture undulates north and south following the solar seasons. Passing over land masses along its way, these clouds release great volumes of rain, giving birth in the process to the rainforests of the Congo, the Amazon, Southeast Asia and Indonesia.

But, as the students recall, the story does not end here. The rising air is not able to descend with the moisture it releases, since more air keeps pushing upward from below. At the top of its vertical ascent—where the upper boundary of the troposphere provides a barrier that keeps it from rising further—it streams laterally to the north and the south, creating a high-altitude *divergence* to complement the sea-level *convergence* below. Drying out on its poleward path, the air that rose out of the ITCZ finally descends—bone dry—thousands of miles away at around 30 degrees north and 30 degrees south of the equator. This descending dry air is responsible for the desert belts—such as the Sahara, Kalahari, Mexican and Australian deserts—that we find at these latitudes. Like "two circulating doughnuts around the earth" (Volk 2003, p 69), this *convection loop*—which stretches from the equator one third of the way toward each pole—creates the wettest and driest regions of our earth! The two cells of this loop are known as the Hadley Cells.[124]

124 Three terms, in particular, highlight what I consider central in this context: *convergence, divergence*, and *convection loop*. Convergence indicates coming together; it often involves compression, compaction, concentration, contraction. Its converse, divergence, is centripetal in nature; it points to dissipation, dispersal, spreading out, expansion. The term convection loop makes evident that a connection exists between these extremes—it indicates that they are not separate phenomena, as such, but rather two poles of one and the same process.

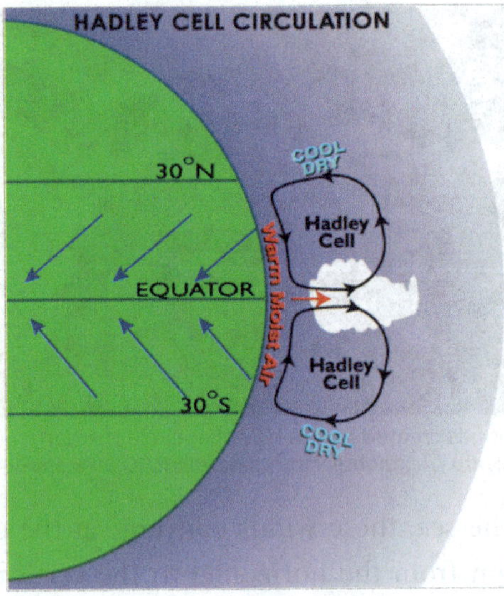

Fig.3.9 The Hadley Cells.(Retrieved from www.newmediastudio.org/DataD iscovery)

Beginning in this way, we rekindle in the students the awareness awakened in their 10th grade meteorology block of the central role that the sun plays, not only in the alternation of summer and winter, of day and night, etc. but also in the overall physiology of the earth. As meteorologist Jack Williams (1997) puts it: "Where the sun shines starts all weather." In their meteorology block they learned how the earth's atmospheric circulation has a direct impact on the global distributions of warmth and precipitation. How the winds and ocean currents redistribute what radiates from the sun, and how water, through its continuous cycling through atmosphere, oceans and land, enables processes that are essential for all life on earth.[125] By starting with the ITCZ, we can pick up those threads from the 10th grade and show how the Hadley Cells create a context for understanding the three primary biomes found on the continent of Africa. (Ahrens 2012; Krump et al. 2009; Volk 2003; Williams 1997)

Africa: A Continental Ecosystem Shaped by Biospherical Rhythms

Even a brief glance at a map of Africa reveals three disparate landscapes[126] that show a powerful contrast between wet and dry that can be attributed to the Hadley convection loops. These are the tropical rain forest, the desert biomes, and the savannas that lie between them.

125 The 10th grade meteorology block usually covers a broad range of topics pertinent to the 11th grade ecology block, including the carbon cycle, greenhouse gases, and fossil fuel consumption.

126 I follow Breckle (2002, p 86), in this usage of the term "landscape."

More than any other continent, Africa is influenced by these convection loops. The contrasts in rainfall across the continent are stunning. Whereas Reunion Island of Madagascar holds the world record for the most rain in a 24-hour period (72 inches!), parts of the Sahara see less than 0.2 inches yearly. On the whole, the wet and dry locations and their seasonality are quite well defined throughout the continent.

Fig. 3.10 Satellite map of Africa.
(Screenshot from NASA's globe software World Wind, Koyos)

The contrast between wet and dry is most acute through the middle of West Africa. The lush Guinea coast runs side by side with the parched Sahara for a stretch of over 1000 miles. This band includes the drought-prone, semi-arid Sahel, that derives its name from an Arabic term meaning border. The northernmost penetration of the ITCZ reaches into the Sahel each year, although the timing and duration of its annual visit varies considerably. Farther north reigns the Sahara, the world's largest desert. In this kingdom of sand and sun, the temperatures are so scorching that the relative humidity drops into the single digits and even the paltry rain that may develop evaporates mostly as "virga" before it reaches the ground. On the narrower, southern end of the continent, the Kalahari Desert forms a much smaller counterpart to the Sahara. Between these zones of dryness, we find Africa's wettest area—the forested valleys and highlands of the Congo Basin. At the heart of equatorial Africa, its climatic regime is very dependable. With the ITCZ pulling wet, thundery weather to the north from March to November and to the south from December to February, it enjoys two wet seasons a year. Even the driest months between June and August normally see at least four inches of rain. With winds typically light, the air is sultry year-round. After explaining the meteorological context for these three biomes—rainforest, deserts and the savannas as transitions between the extremes—each can be explored in more detail (James 1993; Henson 2002).

Tropical Rainforest[127] — The Congo River Basin[128]

Approaching a rainforest from without, from a riverbank or road cut, you are faced with a towering mass of vegetation, solid and impenetrable. Known as the "wall effect," this barrier is roughly 20 feet thick and is compacted with a tangle of light-hungry climbers and vines. To pierce directly through this shield would take hours with a machete, but the initiated know how to follow a stream or animal trail, which allows them to slip behind the wall. Once inside, they suddenly find themselves inside a majestic, cathedral-like space that is often as high as a 17-story building. The thick, pungent scent of unfolding life and decay is striking. Beneath the covering of rotting leaves, a thin layer of rhizome-rich humus can be found. If this is scratched away, a sterile red soil containing a tangled mat of roots is revealed. The eyes require a few moments to adjust to the perpetual twilight, where almost everything appears as a somber shade of green. Known as the "emergent layer," the tallest trees tower above at a height of some 200 feet. The expansive umbrella crown of an emergent tree will sometimes spread over an entire acre. The bases of the medium and large trees frequently send out immense plank buttresses 20–30 feet up the trunk.

The largest trees and their buttress roots can require five minutes to walk around! As Suchantke (2001) puts it: These trees appear to be standing on the earth more than rooted in it. He goes on to describe the inescapable impression that on the forest floor you are actually walking through the root zone of these enormous plants. Almost everything appears root-like.

127 By definition, tropical rainforests contain evergreen trees of at least 98 ft. high (but usually much taller). They are rich in thick-stemmed lianas and epiphytes. Worldwide, their annual rainfall average is 92 inches. Just as important is the relative constancy of rainfall year-round. A tropical rainforest must receive at least four inches of rain each month for two out of three years. Tropical rainforests must have a mean annual temperature of at least 74 degrees Fahrenheit (Newman 1990; Allenby 1999).

128 Next to the Amazon, the Congo River Basin is the world's largest contiguous tropical rain forest. Altogether, the extant tropical rain forests in Africa cover 464 million acres.

Fig. 3.11 Buttress roots. (Valiantrose. CC BY-SA 4.0 International)

Fig. 3.12 The rope-like lianas are often difficult to distinguish from the tree trunks that they seem to merge with.
(JohnJPJr. CC BY-SA 4.0 International)

Fig. 3.13 Rainforest canopy seen from below.
(Mikenorton. CC BY-SA 3.0 Unported)

The earth's surface appears to be high above, up where the light falls in the treetop foliage and is swallowed up before it can penetrate to the ground. Below the canopy, the light is so dim that vegetation is sparse. Maximum illumination comes at noon, when approximately one percent of the light falling on the canopy reaches ground level. At 4 pm, only 0.2 percent will reach the ground. For most visitors to the rainforest, being cut off from the light and immersed in a vegetative world with little space around them gradually leads to the oppressive feeling of being hemmed in. The stifling heat and brooding closeness of the ever-humid atmosphere—which never lets up, night or day—leads to dullness and inertia, which has a smothering effect on all activity and initiative.

Worldwide, tropical rain forests are so similar in structure that the layperson—if limited only to direct sense impressions—would find it very difficult to determine which continent they were on. Even serious students of nature often need some time to discover clues that will tell them of their location. Ultimately, the botanist is able to recognize certain continent-specific species, as is the zoologist. Fascinating, though, is the fact that similar niches on different continents are filled by species that share similar characteristics, but are phylogenetically unrelated (convergent evolution).

When people in the tropics speak of summer and winter, they are referring primarily to the dry and rainy seasons rather than temperature differences. As it turns out, daily temperature variations are actually greater than annual ones! It is,

Fig. 3.14 Tropical rainforest – constant high humidity. (Monotorius)

in fact, only somewhat cooler in the rainy season, and this is due to cloud cover. Almost every afternoon during that time of year, there are torrential downpours lasting several hours, followed by sunshine—although it is not uncommon for rain to continue for several days and nights without a break. Even during the height of the dry season, a newcomer to the tropical rainforest might think it is the wet time of year.

In these forests, humidity is always high, ranging from nearly 80 percent in the dry season to almost 90 percent in the rainy time of year. Seldom penetrated by wind, it is rare for a breeze to be felt on the ground, even as a tropical gale passes overhead. The forest is not unlike a hot house: enclosed and isolated from much of what occurs beyond its upper boundary.

The rainforest canopy, in contrast to the ground level, is rich in flowering trees, vines and other forms of plant life. In the rainforest, the herbaceous layer is not to be found in its usual ground-bound location, but in the form of hanging gardens gathered on branches and vines high above the earth. The combined weight of the epiphytic mass frequently equals one-third of a tree's total weight. (Some limbs become so burdened with the weight of epiphytes that they break off and come crashing down.) During the blooming season, the canopy is abuzz with bats, bees, flies and other pollinators. It is like a meadow elevated high above the earth, for even creatures such as cockroaches, scorpions, mites, springtails and earthworms are found in this treetop world. Although ground dwellers in temperate forests, ants in the rainforest prefer to build their nests in branches that extend out beyond the forest canopy. Some species of termites do the same.[129]

Large troupes of colobus monkeys can be found swinging through the branches. Herbivorous primates are, in fact, characteristic of tropical rainforests worldwide. These grazers of the treetop meadows show parallels both in their physiology and mode of life with ground-bound ruminants.

Fig. 3.15 Ugandan Red Colobus. (Charles J. Sharp. CC BY-SA 4.0 International)

Despite the dense coverage with epiphytes, a clean network of paths exists along the center of certain main branches and limbs. These are treetop game trails kept clear by ongoing use. Most "grazing" mammals of the rainforest are found in the heights.[130] Only a few—like the African duikers or the rare Bong antelope—are found on the ground.

That said, large animals are still completely overshadowed by the plant world in the rainforest. Insects, on the other hand, are overabundant. They exist not only in great numbers, but occur in their largest forms, as well. Forest butterflies and Goliath beetles are examples thereof.

129 There are even epiphytes that grow on other epiphytes. These are called *epiphylls* and include algae, lichen, mosses and liverworts. They may also flourish on rocks and animals. The pinnacle of this life form is the "auto-epiphyte"—a plant that is epiphytic on itself! A few varieties of trees have developed rosettes of litter-trapping leaves that collect debris, into which they grow roots that absorb the nutrients.

130 I use "grazing" here in the loose sense of the word, since those animals that consume primarily the leaves, soft shoots, and fruits of high growing woody plants are normally termed "browsers" by biologists.

Fig. 3.16 Goliath Beetle.
(Fungus Guy. CC BY-SA 4.0 International)

The difference between tropical rainforests and other kinds of forests is huge. Whereas a temperate forest might be dominated by a half-dozen tree species, in a tropical rainforest 500 different tree species can be found in a single hectare.[131] Although data for the huge African rainforest in hard to compile, studies of the smaller rainforest in Costa Rica, which occupies an area less than a quarter the size of Great Britain, give an idea of just how diverse life in a rainforest is. Whereas Costa Rica has roughly 12,000 plant species, Great Britain has less than 3,000. Costa Rica has over 800 bird species, in Great Britain there are fewer than 300. Costa Rica has 137 snake species, Great Britain 3. Costa Rica has 174 species of frogs, toads and salamanders, Great Britain 7. Or, to take another example, in one national park in Peru, more than 1300 butterfly species have been recorded, which is more than three times the number found in all of Europe. The luxuriant life of the earth's rainforests is summed up by the fact that although they cover only three percent of the earth's land surface, they contain almost one third (29%) of the earth's plant biomass.

How is the thriving plant life of the rainforest connected to the earth below? First of all, it is quite surprising to realize that this wealth of organic life thrives above a forest floor that has an extremely thin humus layer and a soil poor in minerals. How is this possible, we ask the students? In contrast to their experiences with the formation of rich garden earth, here the minerals that are released from decomposing detritus do not become part of the soil, but are drawn up immediately by mycorrhizal fungi into the roots and into the organic processes of the tree—

131 Biologists are also fascinated by the fact that although the species of plants and animals found in different rainforests are not closely related, they still look similar and perform similar functions within the web of connections that make up a rainforest (convergence).

above the earth! Were they to go into the soil as such, they would be washed away by daily rainstorms. This supports the impression, reported earlier, that the trees of the rainforest seem to be standing upon the earth more than they are rooted in it. There is no shortage of minerals in the rainforest, but they are not to be found *in* the earth—they remain constantly *above* it in the plant world. For this reason, the slash and burn approach to agriculture that is so common in such biomes has catastrophic long-term effects. After a few years of good crops that grow on the ash-enriched soil that remains behind, the soluble minerals are leached from the soil through rainfall and the depleted remains are baked by the tropical sun.

For much more on these issues, see the section on "Endangered Rainforests" under Additional Topics. (Bockemühl et al. 1978; Bourliere 1973; Breckle 2002; Chapin et al. 2002; Ecologic Development Fund 2018; James 1993; Juniper 2019; Moore 2019; Newman 1990; Perry 1986; Richards 1993; Suchantke 1976, 1982, 2001; Walter 1970)

The Saharan Desert

Marq de Villier and Sheila Hirtle (2002, pp. 1–2) give a vivid description of midday in the Sahara.

> *A group of men, half a dozen in all, are lounging in the meager shade of an acacia in the Sahara Desert somewhere east of Agadez, on the fringe of the sand sea called the Erg de Tenere, in Niger.... It is just after noon. Overhead is the glaring Saharan sun, and the heat is like an iron weight, radiating back from the ground, creating little eddies of convection turbulence, something like sitting on a grill beneath a toaster. An hour earlier the thermometer had climbed past 105 degrees to 115 and then 120, and the steel of the battered Toyota Land cruiser is too hot to touch, the metal pinging in protest.*
>
> *The conversation has wound down, and they are sitting in silence, all talk dried up in the superheated and desiccated air. The breeze is far from a gale, but enough to stir the desert sand, fine as flour. It drifts through the camp and overhead, turning the sky a hazy yellow and blurring the inescapable sun. They have parked their vehicle to windward, but it does nothing to hinder the sand; a tin plate placed on the ground is covered in seconds with a thin gritty film. Sand fills the ears and the nose and the reddening eyes and infiltrates the clothing, drifting sand in every crack and crevice: sand in the tea glasses, sand in the food, a dismaying grit on*

the teeth. Visibility is minimal, perhaps 30 feet. They sit hunched, merely waiting. Wantam, the leader, has his head on his knee; he seems to be barely breathing.

Out of this sulfurous, swirling obscurity loom three men on camels, plodding steadily through the veil of sand. They pass within a few yards of the camp, but no one says anything, or raises a hand in greeting.... They move through the camp silently, the camel's feet making no sound on the soft sand.... Where are they going, these three deep-desert Tuareg? Where have they come from? On what errand? And how, in this trackless land of no horizons, in this minuscule visible universe of maybe 20 paces do they know their path? They are heading north, but north is nothing but weeks of difficult traveling to nowhere. To the west is the stony massif called Air, but to the north and east—nothing.

The Sahara is the earth's largest and most austere desert. Outsiders who visit the Sahara often speak of a sea without water, of a land of terrible thirst, of a wilderness of sand and stone and barren rock. Indeed, visitors to the Sahara may travel for many days without discernible landmarks to register their progress. You can plod through the Sahara apparently forever, in shimmering heat from sunrise to sunset, through a landscape of countless stony valleys and plains, of rock slabs and sheets of gravel, each one as featureless as the one before it. For outsiders this landscape can lead to despair. To the nomads of the desert, however, the stony heart of the Sahara is not an empty sea, but a place of openness. One resident of the Sahara, Ibrahim Salam Bouseif, explains how it appears to him:

This may seem like the end of the earth to you, but it is not for us....In the desert we are free; there are no restrictions. We can see forever. We can cast our eyes to the horizon, and all we can see is sand and sky. In the desert our spirits can fly. As the desert saying goes, Allah has removed all surplus human and animal life from the desert so that there might be one place for him to walk in peace... and so the Great Sahara is called "the Garden of Allah." (de Villiers and Hirtle 2002, p 5).

Indeed, if you fly over the Sahara in daylight, for example from Marrakech to N'Djamena in Chad, you can stare out the airplane window for six hours without seeing a living thing on the ground. Even when flying at relatively low altitudes, the wandering human tracks and game trails that mark most arid landscapes are almost entirely absent. What can be seen is a color palette: massive stretches of caramel and beige sand-seas blending into the grays and duns of the hamadas[132] and the dusty-silver salt pans. In places, sharp spikes of black basalt and patches of iron appear where mountains arise. The few isolated oases you may pass over appear dun against dun—unless the sun is just at the right angle, which allows you to spot the dusty green of a date grove. But even the largest of groves looks tiny and as though clinging to life in the midst of a desert that is overtaking it—which indeed it is in many cases.

The common image of the Sahara among outsiders is that of sand and a few scattered oases. In fact, only about 15 percent of the Sahara is actually sand. Over four-fifths of the desert consists of mountains and massive plains of gravel and rock. Rock and gravel is the most characteristic Saharan landscape.

Fig. 3.17 Reg in Adrar (Mauritania).
(Ji-Elle; GNU Free Documentation License)

Known as "regs," these landscapes are plain beyond belief: colorless and totally lacking in memorable features. It is possible to travel from morning to night for two whole days without viewing a tree, a shrub, a blade of grass or even a boulder of any size, not to mention a mountain or mesa, a valley or a well. What you experience is merely a flat gravel plain![133]

132 Hamadas is bare rock with isolated boulders.

133 Of course, those who make the desert their home and who are intimately attuned to its subtle variations, the Tuareg, have many different names for the many fine subdivisions of desert that escape the normal traveler's notice.

Fig. 3.18 Sahara Desert. (Flickr; Author: Wonker)

Despite being only fifteen percent sand, every Saharan country has massive dunes and seas of sand that can reach for tens, even hundreds, of miles. The dunes move very quickly.

Fig. 3.19 Saharan sand dune. (PierreHaut. CC BY-SA 4.0 International)

What today is only barren soil, thorns, broken stone, gravel... can become in only a few weeks a full-fledged, honey-colored dune that has been sculpted by the wind, with a knife-edge to the leeward and rising to a height of 50 to a 100 feet. Dunes range in size from a small pile of sand to a giant mountain of 1,600 feet or more. The shifting of larger dunes creates a hissing sound. In the full glare of the sun—with the hot winds blowing and the temperature moving past 120 degrees—

the whole landscape seems to migrate. In a high wind, the dunes smoke and a low-pitched drumming, caused by the avalanching of billions of tiny sand grains, is audible. The tiny sand grains skitter across the landscape in little hops; as they skip, they knock into other tiny grains and bring them into movement (saltation). Dunes not only advance, but they also diminish, often vanishing as quickly as they appeared. Even a massive dune can turn into a flat sand sheet within a month.

Fed by underground rivers and basins, fertile oases that grow dates, corn and fruits make up only about 800 square miles of the vast Saharan desert. Some shrubs and trees can be found on the edges of dry valleys (wadis). These phreatophytes[134] have huge root systems and meager above-ground sections. With their long roots they can reach down through several meters of overlying dry soil to the ground water below. The roots of the famous Tree of Tenere[135]—to use an exceptional example—were found at a depth of 115 feet.

Fig. 3.20 Tree of Ténéré. (Source: CC BY-SA 2.0)

Except for the rainless central Sahara, where it is possible to travel for days without meeting the slightest trace of plant life, most areas have at least a brief wet

134 Phreatophytes are deep-rooted plants that obtain water from a permanent ground supply or from the water table.

135 It is shocking for the students to hear the story of how this solitary acacia tree—the only tree within 150 miles—came to an end. Passed by salt caravans on a regular basis, the tree had an almost mystical status and every traveler knew—no matter how desperate they were for firewood—that they dare not cut it down. But alas, after a road of sorts had been built near it, one day this three- to four-hundred-year-old tree was knocked down and destroyed by a drunken truck driver. The place where it had once grown is still marked on maps and its trunk was saved and lies in the natural history museum in Niomees. The tree also lives on as a metaphor for the tenacity and toughness of the Tuareg nomads who live in this region.

season. During these short periods of rain, the ephemerals unfold. They are able to germinate within 48 hours of even the slightest rainfall, and sow their seeds as early as two weeks later. Some ephemerals form occasional pastures highly valued by the nomads. Known as "acheb" when they spring up after the rains, these "pastures" are far from dense, with individual grass plants up to 50 yards apart.

Xerophytes, on the other hand, survive long waterless periods in a state of reduced activity. Their leaves are often contracted to spines or scales. These plants frequently appear as "pitiful looking cripples" (Breckle 2002), although they may live for 100 years or more. Many branches may die, but enough survive to allow further growth when the rain returns. A special group is made up of succulents. These species store water and use it sparingly during times of drought. Three groups can be distinguished: One has succulent leaves, another succulent stems, and a third possesses succulent roots.

Fig. 3.21 Rose of Jerico *[Anastatica].*
(Ji-Elle. GNU Free Documentation License.)

One well known desert dweller is the Rose of Jerico, a member of the mustard family. The small grayish plant curls its branches and seedpods inward during the dry season, forming a ball that will open again only when moistened. It can survive for years in that form, although it is sometimes uprooted and blown about by the wind like a tumbleweed. If still rooted when moistened, it spreads into a green plant as much as 1 foot wide and bears tiny white flowers.

Many xerophytes maintain a low level of sap concentration and can survive up to a year without water by means of Crassulacean Acid Metabolism (CAM). They open their stomata at night when water loss is minimal and take up carbon dioxide, which is then incorporated into organic acids. During the day, when the stomata are closed, the carbon dioxide can be used in a "delayed" photosynthesis process and water loss is avoided.

Fig. 3.22 Salt-tolerant, drought resistant shrubs. Taghit, Bechar, Algeria. (LBM1948. CC BY-SA 4.0)

Halophytes also constitute an important group of Saharan plants. They colonize saline soils. These salt-tolerant plants can be found on the periphery of former lakes, the lower reaches of rivers and salt pans.

Fig. 3.23 *Welwitschia mirabilis*, Naukluft, Namibia. (Sara&Joachim on Flickr. CC BY-SA 2.0 Generic)

One desert dweller that students find particularly fascinating is *Welwitschia*. An adult *Welwitschia* consists of only two leaves, a stem, and roots. Its two leaves are permanent—they are the original leaves from when the plant was a seedling. They just keep growing and are never shed. The leaves are leathery, lie on the ground, and become tattered and torn to ribbons with age. The stem is woody, hollowed-out, and sturdy. *Welwitschia's* lifespan is estimated to be 400 to 1500 years, with growth occurring only during the summer months.

Fig. 3.24 Almost ripe cones of a female Welwitschia plant.
(Hans Hillewaert. CC BY-SA 4.0 International)

The stamen-forming flowers produce salmon-colored, cone-like structures, while the seed-forming cones are blue-green, larger and more tapering. The seeds remain viable for a number of years, germinating only if fairly heavy rain is spread over a period of several days. Since this rarely occurs, in many colonies numerous plants will be the same age. *Welwitschia* is dependent on deep groundwater; hence its occurrence in dry riverbeds and watercourses, for which it needs a deep taproot. It also has many lateral roots just below the surface that take advantage of surface moisture. Antelope and rhinos chew the leaves for juice during times of drought and spit out the tough fibers. This does not damage the plant since it simply grows out again from its meristem tissue.

To live in the desert is to live "in spite of it" (Suchantke 2001, p 52). Animals do this in different ways, such as hiding away during the heat of the day or by reducing the need to drink through minimal evaporation. The latter is achieved by the desert fox, the gazelle and certain other antelope, which fulfill their water needs through their food.

Fig. 3.25 Dromedaries in the Sahara, Morocco.
(Bikehiker75. CC BY-SA 4.0 International)

The dromedary is able to play such a central role in the desert because its blood does not thicken, even under conditions of extreme water loss. It can tolerate a loss of up to 40 percent of its bodily fluids without harm![136] Another master of desert dryness is the sandgrouse (Family Pteroclidae). Often required to travel long distances in search of water for their young, these creatures transport it on the return trip in their saturated under-feathers.

Fig. 3.26 A male Addax. (Haytem93. CC BY-SA 4.0 International)

136 This is the equivalent of 25 percent of its body weight. In humans, a 10 percent decline in body weight through water loss is fatal.

Antelope are often called the most exhilarating of all desert animals. Of the three species still found in the Sahara, the most impressive is the fleet-footed addax—the desert's largest wild mammal. Not much more than three feet tall at the shoulder, they are believed to have the capacity to sense desert grasses and bush at a distance of 50 miles or more. Found only in isolated spots today, they have been seen ninety miles from the nearest water source moving over rough terrain at speeds of over 40 miles per hour.

Most desert creatures, however, are small, live in holes or burrows, and emerge only at night. Small rodents and insects are the most common. The ferocious, two-pound fennec fox feeds on such rodents and derives its moisture from them. Its fur-soled feet insulate it from the burning sand and its huge ears radiate heat from its body.

Snakes populate many parts of the Sahara, as well. The Saharan horned viper and the sand viper are the most common species. While travelers tend to exaggerate the hazards of snakes, scorpions are a more pervasive problem.

Fig. 3.27 Saharan Death Stalker Scorpion. (מינוזיג)
CC BY-SA 4.0 International)

They seem to be everywhere and are conspicuous in almost all chronicles of Saharan travel. Saharan scorpions can reach six inches in length. Their sting can cause paralysis—usually temporary—and even death. They are considered among the oldest "living fossils" on earth. Which is to say, they have remained unchanged amidst untold transformations in their environment and in the creatures around them.

Stepping back and reflecting in small groups on a description of desert life something like the one above, the students are not completely surprised, but fascinated nonetheless by the significance that water plays in the unfolding of life on earth, and by the amazing capacity of some organisms to deal with the lack of it. In some ways, the unchanging scorpion could be seen as a symbol for the absence of water's transformative powers in the desert.

The huge differences between desert and rainforest—both of which exist on the same continent—is amazing. In contrast to the shadow-casting rainforest—which is water-rich and soil-mineral poor—the great desert is blanketed in earthly minerals, bone dry and always exposed to the baking heat of the tropical sun. What the one landscape has in overabundance, is missing almost entirely in the other. But, not surprisingly, located between these two extremes we find yet another world with its own set of unique characteristics. (Braeckle 2002; de Villiers & Hirtle 2002; Odum & Barrett 2005; Suchantke 2001; Van Jaarsveld 2000)

The African Savanna

One visitor to the African savanna observed the following:

> *Grass waves gently in the wind on a rolling plain, studded with acacia trees. Lions lie, creamy-white bellies exposed to the sun, eyes tightly shut. Way in the distance, where the silhouettes of delicate lace-canopied acacia trees break the horizon, Thomson gazelles graze, rich chestnut coats broken up by black and white stripes sharp in the sun…There are massive herds of white-bearded wildebeest and smaller groups of wise-looking topi and kongoni. Looking every bit like nursery toys, tight knots of zebra quarrel and bicker, while in the woodland glades, herds of wide-eyed impala and rather stately waterbuck are found. In quiet corners lurk pairs of dik-dik and family groups of oribi. A frisson of danger is provided by carnivores like leopard, cheetah and lion, plus serval, caracal, wild cat and genet. Vibrant flashes of color light up the scenery from a variety of birds. There are the exotic electric blues and purples of lilac-breasted rollers and the vivid greens and orange of the Fischer's lovebirds. Lovebirds like company and, as they dart off in panic, the tight, chattering flocks catch the light like bright handfuls of jewels carelessly tossed into the air.* (Morris et al. 2002, p. 200)

Visitors wax eloquent about the sight of thousands of grazing animals spread out over an African savanna. No other continent contains such an abundance of

Fig. 3.28 Rothschild's giraffes in Murchison Falls National Park. (Uganda) (DrexRockman)

Fig. 3.29 Serengeti savanna plains, Tanzania. (Gary. CC BY-SA 2.0 Generic)

large mammals, and the majority of these live in the savanna. Most of these animals are part of large herds that migrate yearly. Within the great variety of grazing animals, the species all feed differently and mingle peacefully.

Elephants and buffalo pull at long grasses and trample them. This stimulates the growth of finer grasses that zebra and wildebeest seek. First come the zebras, which eat the drier and rougher grasses. They are followed by the wildebeest—strewn out in herds of up to 10,000—that nibble the grass to below ankle height. The really short grass is then grazed by warthogs and gazelles, while the leftovers are nibbled away by Egyptian geese. Such complete grazing of the savanna provides much less fuel for the fires of the dry season than do areas covered with dry, uncropped grass. Thus no fires can arise that are hot enough to destroy grass roots, soil fauna, or thick-barked acacias.

Fig. 3.30 Termite mounds in Namibia.
(Olga Ernst. CC BY-SA 4.0 International)

Amazingly, the greatest consumers of grass are among the tiniest: the termites. In the savanna of northern Kenya, for example, more than 80 percent of net primary production is consumed by termites. In the Congo savanna about 30 percent of the soil surface is covered by termite mounds. Each colony may contain seven million or more individuals. These consume up to one-third of the total yearly production of dead wood, grass and leaves. They are able to digest cellulose through their associations with endosymbiotic enzyme-producing fungi, bacteria and protozoa. Although their excavation can accelerate soil erosion by clearing off plant cover, the mounds are often very fertile due to the large amounts of organic matter and vital plant nutrients they contain.

In areas where tree growth is denser, patterns such as the following can be observed:

> *When elephants frequent any portion of forest for any length of time, they are invariably followed by the scavenging wild pig, at times by baboon, which take advantage of the roots and bulbs displaced by the great animals, and which are not above searching the droppings for food. The disturbance to the soil caused by the elephant, the wild pig and the baboon, brings about soil improvement, and stimulates many dormant seeds to germinate. Fruits passed though the animals are cleaned of their outer coverings and fall into improved germinating beds.* (Golley 1993, p. 11)

Fig. 3.31 Lion male in Masai Mara National Park.
(Byrdyak. CC BY-SA 4.0 International)

In many areas, predators—lions, leopards, cheetahs, spotted hyenas, jackals—play a decisive role. Although they are quantitatively negligible, they keep the grazer population within limits, beyond which hunger, sickness and phenomena connected with over-grazing set in. In this way, they help provide ecological stability. Birds of prey can be seen in a similar light.

With their songs and calls and the splendor of their colors, birds animate the landscape on a level equal to the mammal population. The radiant colors of glossy starlings, bee eaters, sunbirds, and incandescent kingfishers, contrast greatly

Fig. 3.32 Red-billed Quelea. (CC BY-SA 2.0 Generic)

with the tawny savanna background. Birdsong is said to reach new heights in the savannas of Africa. Like the mammals, Africa's bird life is not concentrated in the forest but in the savanna.

Savanna vegetation is characterized by an open overstory of trees and a ground layer of grasses. Although these two vegetation types normally exclude each other, in savannas they combine in a harmonious unity. This is possible because just enough water enters the soil during the rainy season to keep a limited number of trees alive during the dry periods. During the drought season, they either reduce transpiration to a minimum—which the acacia, for example, achieves by folding up its thin leaflets during the heat of the day—or shed their leaves. Loosely spaced trees allow plenty of sunlight for the grasses beneath them, which grow abundantly. Trees and grasses do not compete for water due to their fundamentally different root systems. Whereas the very finely branched, intensive root system of grasses permeate only a small volume of soil, the extensive root system of trees extends far into the earth in all directions, thereby penetrating a much larger soil volume, but doing so much less densely. This balance of trees and grasses shifts if rain levels increase and more water remains in the soil during dry periods. Then the trees multiply, savanna turns to dry forest and the grasses recede for lack of light. On the other hand, when the rainy season brings too little water to sustain soil moisture, trees can no longer survive and the savanna becomes grassland.

Fig. 3.33 Umbrella Acacia amid grasses. (Kufundisha. CC BY-SA 4.0 International)

The two characteristic trees of the African savanna, the umbrella acacia (*Acacia tortilis*) and the baobab (*Adansonia digitat*) present a striking contrast (Figs. 3.33 and 3.34). The thickly leaved crown of the acacia provides animals with needed shade during midday heat. The slight, powerless appearing trunk is dominated by the spreading crown.

Fig. 3.34 Baobab Tree. (Licensed under the CC BY-SA 3.0 Unported)

The baobab gives the reverse impression. Its massive trunk is topped off by tangled branches. The trunk consists of soft wood that stores vast amounts of water during the dry season. This becomes a water source for elephants: They tear out large chunks of the baobab trunk with their tusks and suck on them.

Fires are as integral to the savanna as its animals, grasses and trees. They act upon the landscape in a manner similar to grazing ungulates. Both help keep the savanna open, thus preventing it from becoming an impenetrable tangle of thorn bushes. Not only lightening, but also human beings have also been decisive in shaping the savanna through fire and grazing. The savanna is home to pastoral peoples, who migrate cyclically with their cattle, following the seasonal rhythm of dry and rainy periods. These pastoral nomads are juxtaposed with farmers. The herding peoples prefer the drier regions of the savanna, where grassland predominates, while the farming peoples have settled more in areas where the savanna meets the forest.

Of Africa's ecosystems, the savanna is clearly the one most strongly influenced by seasonal change. In the rainforest it is common to find trees on which some branches are bare, some are budding, some are shedding their leaves and yet others are in full foliage: All four seasons are present simultaneously on the same tree! The desert, by contrast, is so lacking in vegetation that growth cycles are absent. In both desert and rainforest we find, moreover, that daily variations in temperature (in the desert) and rainfall (in the rainforest) are much more extreme than seasonal variations. In the Sahara, for example, the temperature swings from day to night are greater than those between summer and winter.

Seasons in the savanna represent milder versions of the extremes found in the desert and rainforest that border upon it. The savanna pulses rhythmically between one pole and the other without falling completely into either extreme. As Suchantke points out, this rhythm is analogous to breathing. During the dry season, life in the savanna "breathes in": The herds have migrated, many birds have moved on, reptiles and amphibians have gone underground, grasses have withdrawn, and trees have shed their leaves. The air is dusty and smells of smoke from the many brush fires that have charred the land with black patches. When the summer—the rainy season—returns, the landscape "breathes out" again: Grasses spring up, bushes and trees burst into leaf, acacias blossom, dense swarms of termites stream out of the ground, tadpoles and small fish fill short-lived puddles, the whole animal world teems with newborn animals. For humans, it is the time of sowing and planting. (Breckle & Walter 2002; Gichohi et al. 1996; Golley 1993; Morris et al. 2002; Suchantke 2001; Walter 1970)

Summing Up

The preceding sketch of Africa's three primary ecosystems reveals a dynamic polarity between the verdant life-pole of the rainforest and the largely lifeless expanses of the desert.[137] The savanna represents a dynamic balance between these two extremes, alternating rhythmically between wet and dry, between unfolding and withdrawing, in the course of the year. This mediating position is evident from its geographic location, lying as it does between deserts in the north and south and the equator-straddling rainforest at the continent's center.

In the rainforest we find an immense, upward-sweeping gesture as the root-sphere rises above the earth's surface, gathers into column-like trunks that unfold in the heights, umbrella-like and adorned with radiant epiphytes. It is as if the

137 In the above I have considered neither Africa's alpine landscapes, nor two regions at the extreme edge of the continent—the Mediterranean zone in the north, and the southern tip of the Cape Province. These areas are atypical for Africa, as they are the only two regions affected by recent rock folding, and furthermore, in both cases, large African animals have been mostly eliminated (Suchantke 2001).

atmospheric flow from the depths below to the airy heights above—which the Hadley Cell exhibits on a macro scale—is manifest here in the flow of minerals and water from the earth below up into the spreading canopy.

In the desert, on the other hand, all plant life is held back. The greatest part of the plant lives under the surface (phreatophytes), or is contracted into seeds, tubers or bulbs for most of its life (ephemerals). Many plants that maintain an above-the-surface presence, like the succulents, "breathe" only in a very halting manner through the two phases of Crassulacean Acid Metabolism. In contrast to the rainforest's out-breathing of water—up from the earth below, through the plant and into the atmosphere (transpiration)—desert plants hold the water back, keeping it within by means of closed stomata and contracted forms. Here, where the Hadley convection cell descends from above—hot and dry—all flow from below upward is suppressed.

The savanna lies between these extremes. Its "gestalt" is more temporal in nature. It breathes out in the wet seasons, breathes in when it is dry. This in- and out-breathing moves across the African continent—from north to south and back again—in a rhythmical pattern. This is revealed not only in the plants, but also in the migration of the savanna's animals and nomadic residents. When the northern savanna breathes in, the southern savanna unfolds, and vice versa, as the year progresses. It is here, in this rhythmical middle of the continent, where the migrating herds of large mammals are at home.

This continent-wide polarity between wet and dry, between expansion and contraction, between oppressive light and expansive life—mediated as it is by the rhythm of the seasons in the savannas—usually reminds the students of other polarities we find in nature—such as day and night, summer and winter—and in our own bodily organization as we discussed in our 10th grade human biology block.[138]

> *A nerve cell loses the capacity to divide almost immediately after birth....*
> *The greater part of the eye manifests practically no "life" in this sense.*
> *If one pursues this line of thought, which can be mentioned only briefly*

138 If time is available and the students show an interest in pursuing this perspective more extensively, it is very interesting to see how a qualitative similarity can be found between the desert–rainforest polarity—mediated by savannas—and the nerve-sense system–metabolic system polarity—meditated by the rhythmical system—in the human organism. Rather than expand on this topic here, I refer the reader to a valuable resource in such matters—Husemann/Wolff's An Anthroposophical Approach to Medicine—with a longish citation that illustrates this relationship which text follows.

here, one must conclude that in the nerve-sense system the opposite of life, namely, death, is active. Portions of these organs are already almost dead during life and are able to fulfill their function just because of this. Thus, in the lens of the eye, one finds no blood and a minimal metabolism. The same is true of the white matter of the brain. It is obvious that humans cannot live only from organs so constituted, that life itself must be connected with other organs....

In fact, there are organs in which the opposite tendency is dominant.... There are organs whose main function is the conversion and transformation of substances. The most important organ of metabolism is the liver, in which transformation of all kinds of substance takes place continuously. Such transformations—intense metabolic processes—are not only present in the muscles and liver, but also in the intestines, kidneys, and blood. These organs can be seen as the foci of metabolic transformation. Together with the process they manifest, they can be called the metabolic-limb system. In fact, these organs show a clear relationship to life; the words "life" and "liver" are etymologically related in many languages. If one follows the functioning of the liver in all its manifestations, once can see that it is really a principal organ of life. Life manifests and expresses itself in the transformation of substances.

Looking at the nerve-sense system and the metabolic system as polar opposites, one discovers not only that their chief organs are spatially separated, but also that their described tendencies and attendant phenomena always oppose one another.... In reality, the two opposing systems described are not active simultaneously, but alternate constantly and rhythmically. During the day, man is awake and "lives" from the forces of the nerve-sense system. At night, during sleep, the actual life processes unfold; a general warming, enlivening, and stimulation of metabolic processes occurs. At this time there is a "switching off" of nerve-sense activity; the human being sleeps. We see that the rhythm of waking and sleeping expresses the alternation of dominance by one or the other polar system, by nerve-sense activity or metabolic activity.

The alternation or rhythm is found again in virtually archetypal manifestation in the heart and lungs. Their rhythmic activity is so obvious

> *that these two organs can be viewed as representatives of the rhythmic system. The heart alternates uninterruptedly between contraction and expansions, and one can see in contracting and tensing the opposite of expanding and relaxing. The dynamics described are directly apparent here. The same is true of the lungs with their continual alternation between inhalation and exhalation.... In contraction lives the dynamic of the nerve system, in expansion, the dynamic of the metabolic system.* (Husemann/Wolff 1982, pp. 7–10)

Ecological Interdependencies

Digging into More Detail with a Few Examples

The students are often interested to hear how Frederic Clements—the most influential American ecologist in the first half of the 20th century—emphasized that mature ecosystems (climax communities) were inseparably connected with their climate and could be viewed as complex organisms. The term "ecosystem" was first coined in 1935 by the British botanist Arthur Tansley in an effort to emphasize that an environment and all living organisms need to be seen as a single interactive whole.

After expanding out into such broad perspectives, it makes sense to contract again into some of the more detailed aspects of life within an ecosystem. Although it helps keep such topics alive if the teacher seeks out new instances from year to year. I will begin with one particular example of the amazing interdependency that exists between various organisms and their environment that never fails to interest the students (and the teacher). It is a topic that can be approached with a riddle. (Clements 1916, 1935; Tansley 1935)

The Australian Crisis and the Creatures That Solved It

Mid-20th century biologists were faced with the interesting question of why two climatically similar regions, the savannas of Africa and much of Australia, had a very different relationship to the dung (excrement) produced by the hoofed animals that inhabited those regions. Whereas in Africa freshly deposited dung disappeared rapidly and served to increase the fertility of the soil, in Australia dung from hoofed animals created great problems. Each of the roughly 25 million Australian cows produced around 10 large manure piles (pasture patties, meadow muffins) daily. These remained lying on the ground for months—even years—and covered thousands of acres of valuable grazing land. It was estimated that for each cow approximately 400 square meters of pasture was lost. An additional downside of the untransformed dung was that animals avoided the grass around the patties because it was too nitrogen rich. As a prerequisite for any meaningful discussion of this situation, the students must first learn about an important difference between these two ecosystems: Before Europeans intervened, the only placental mammals in Australia were bats, rats and mice. Hoofed animals, dogs, cats, etc., were first introduced to the land by Europeans in recent centuries.

With this as background, we can let the students ponder what might be the cause of the dung problem. They will not have much success, however, until they become familiar with a critical component of all ecosystems worldwide (with the exception of Antarctica) that will provide the solution to this riddle. To introduce this pivotal ecological element, I refer back to Africa to see how the patties there get off the surface and into the earth.

There are many variations on the theme that follows, and teachers must limit themselves to a few, of course. My favorite starter-example begins with elephants. I ask the students to imagine a herd of a hundred or so dust-covered elephants ambling slowly along grazing on fresh green grass, nimbly pulling up tufts that have just sprung up, and stuffing them into their mouths. Each elephant consumes hundreds of pounds of grass and twigs in this way daily. As an outcome of this process, they frequently leave behind basketball size droppings (dung piles), a trail of which the herd leaves behind as it moves along.

If you go out right after dark to where the herd passed that day and wait, you will soon hear a low humming sound. It is the sound of hundreds and thousands of dung beetles barreling in—all of them making a beeline for the dung piles that they

Fig. 3.35 Namibia – Dung Beetle feast.
(Duwwel. Public domain)

are able to locate through their well-developed sense of smell. Berndt Heinrich and fellow researchers from UCLA once put out a half-liter of fresh elephant dung and within 15 minutes they collected and counted 3,800 beetles coming to the pile. The total weight of the beetles was greater than that of the dung they had put out!

The beetles range in size from barely larger than a grain of rice to the well-known elephant dung beetle (*Heliocopris*) that Heinrich describes as follows: "This sparrow-sized beetle is built like a tank, flies like a hawk, and tunnels through hard-packed soil like a bulldozer (Heinrich 2012, p. 131)."

Every night, the dung beetles deal with the dung left behind in one of three basic ways. The so-called "dwellers" burrow into the patty and lay their eggs in it. They are generally small and remain living on site until the dung is exhausted. The sparrow-sized *Heliocopris* is a classic "tunneler." When they come flying into an elephant dung pile, they crash-land, fold under their large wings, and lumber over to the pile. After picking a location, they tunnel straight down into the ground under the pile. They usually appear in pairs, so that when the tunnel is complete, the male remains near the top of the tunnel and brings dung down to the female. The female forms the loose dung into a round or pear-shaped brood ball. One egg is then laid in each ball, which contains all the food and water a single larva needs for its development.

The third way of dealing with the dung is undertaken by the "rollers," who come in all sizes. One classic roller, the species *Scarabaeus laevistriatus*, comes flying in at 30 kilometers per hour and hits the ground running.

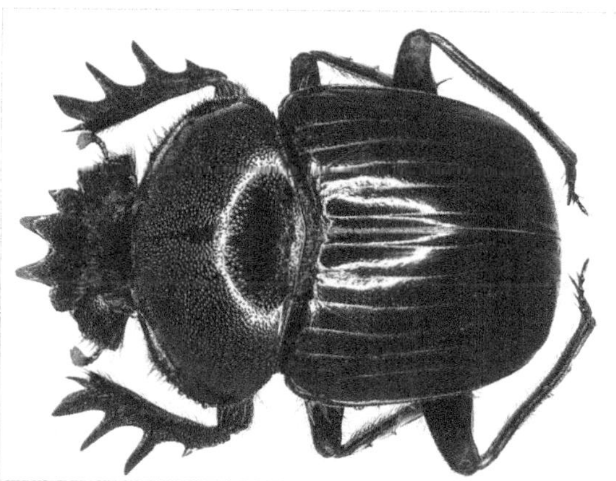

Fig. 3.36 *Scarabaeus laevistriatus.*
(Udo Schmidt. CC BY-SA 2.0 Generic)

Often in male and female pairs, they arrive at dusk and inspect the dung with their antennae before cutting into it with the rake-like prongs on their front legs and with the shovel-like extensions on the front of their head. They pry off chunks of dung in this way and pat additional dung onto the original pieces with their front legs. Within ten minutes to half an hour, they sculpt out an almost perfect sphere ranging in size from a golf ball to a baseball.

Fig. 3.37 Dung Beetle rolling a dung ball.
(Bernard DUPONT. CC BY-SA 2.0 Generic)

Although the details vary between species, typically the male and female first encounter each other either at the dung dropping or while the male is making or has just completed a dung ball. In most instances the male then places its long hind legs—handstand-like—on the ball, with its front legs on the ground. It starts walking backward, pushing off with its front legs while kicking the ball along with its hind ones. The female with often climb on the ball and "hitchhike" while it is rolled away to a spot with soil soft enough for burying the dung. The male buries the dung ball by excavating directly beneath it so that it gradually sinks into the ground. The female, sitting on top of the ball, sinks with it. One egg is then laid in what is now called the "brood ball." Some species will go back to the pile to create another ball, and then another, placing them on top of each other like a sleeve of tennis balls, each containing one egg. The hole will then be closed off and the larvae left to hatch, feed, and develop on their own.

What does all this dung beetle activity mean for the ecosystems where they reside? One central effect is obvious by now: They remove masses of manure from the surface that would otherwise accumulate, thereby choking out plants and limiting the populations of animals that live from the surface vegetation. But they

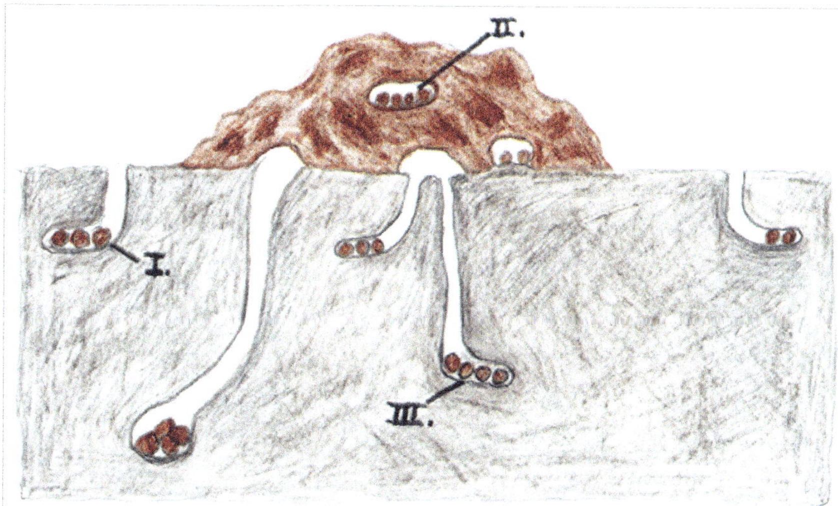

Fig. 3.38 Different forms of brood ball storage by rollers (I), dwellers (II), and tunnellers (III). (Modified after Grzimek and Makweti)

do even more. Their activity also loosens, aerates and nourishes the soil, as well as retards the spread of parasites. It also facilitates the dispersal of seeds that are present in animal dung. Another very visible and important effect is the great reduction in the number of very bothersome flies that would otherwise breed in the dung. And, as some researchers have pointed out, the world would be a much smellier place without the efforts of these amazing beetles!

Although our dung beetle example has been limited to Africa, when we remind the students that dung beetles reside everywhere on earth except Antarctica, and come in more than 5000 variations (species), then it becomes clear that they play a huge role in the life of the earth as a whole. We can now return to Australia and ask the students what they think created the huge excess-excrement problems there. The answer is now obvious, although there are several loose ends that still need illuminating. Namely, if Africa has over 2000 varieties of dung beetles, why doesn't Australia? Biologists who were studying this problem soon realized that Australia actually had indigenous dung beetles too, but these were only able to deal with the extremely hard, dry, and pill-like dung balls produced by the Australia mammal population, none of which were hoofed animals. The Australian dung beetles were not able to work with the soft, watery excrement produced by the hoofed animals that had been brought to the continent by Europeans.

Once this became clear, the followup question for ecologists was: What can we do about this? The students are quick to realize that the careful importation of "fitting" dung beetles might be the solution. And, indeed, this became the central idea behind a 20-year project—The Australian Dung Beetle Project—led by the Hungarian entomologist G.F. Bornemissza. Careful is a pivotal word in this context because Bornemissza knew that introducing non-indigenous species is always potentially dangerous. After searching in 32 countries for the most suitable species to meet the Australian problem, the chosen species had to be quarantined to ensure that they did not carry parasites that could become pests. Eventually, 55 dung beetle species were introduced into Australia by the project. Each year after their introduction, the beetles expanded further and further into extensive new areas of N.E. Australia, making the project an immense success. Cow pies were now disappearing into the earth in two or three days, and the bush fly, which had plagued the land, was now brought under control. In the final report on the Dung Beetle Project in 2001–2002, it was stated: "The impact of this on soil, water and pasture health, and on control of pest flies is undoubtedly worth many millions of dollars a year." In fact, the success of the project is said to be the reason why Australians can now enjoy an outdoor café culture. Up until then, bush flies had been such a pest that it was impossible to have outdoor dining without a designated area enclosed by fly-wire. Another consequence is that the classic "Australian salute"—a flick of the hand used by generations of Australians to shoo away flies—is becoming a "dwindling gesture." (Cribb 2006; Grzimek 1979; Heinrich 2012; Heinrich & Batholemew 1979; Makweti 2020; Schad 1985q)

Hoofed Animals and Their Predators

After exploring the synergy that lives between hoofed animals, savanna vegetation and dung beetles in considerable detail, it is interesting to look at another relationship that plays a formative role both in the life of the millions of hoofed animals that migrate across the African savannas, and in the vegetation found there. This is the connection between the hoofed animals and their predators. Although the students usually find the six large predators (lions, leopards, cheetahs, hyenas, wild dogs, and jackals) quite interesting in themselves, they often have a strong antipathetic reaction when it comes to the relationship of the predators to their prey (wildebeest, zebras, gazelles, impalas, giraffes, buffalo, etc.). This is understandable on an individual basis, but from a larger perspective—from the perspective of the entire ecosystem—predation by large carnivores takes on great importance.

It is therefore valuable for the students to hear what George Schaller—one of the world's preeminent field biologists—has to say about this:

> *Unless they are controlled in some way, hoofed animals tend to increase up to and even beyond the capacity of the habitat to support them on a long-term basis. Disease then becomes rife, and, with the food supply exhausted, many may die of starvation. One of the most important functions of predation is to keep prey species within the carrying capacity of the land, below a level at which disease and starvation can take effect. By taking a percentage of healthy, reproducing animals, predators help to prevent population explosions and the resulting crashes by dampening the rate of increase in the prey. They contribute to the maintenance of an equilibrium within the limits imposed by the environment. The lion and other predators are in integral part of the ecological community and must be present if the prey animals are to survive in all their vigor and abundance.... All species are bound to each other in a universe of relationships and maintain a tenuous harmony, each dependent on the other for survival.* (Schaller 1983, p 166)

The Effects of Grazing

We find another interesting variation on this theme when we turn our attention to the effects that the grazing ungulates have on the vegetation that they are consuming. For over two decades, a team of researchers led by Syracuse University's Sam McNaughton investigated the complex interrelations between plants and various aspects of their environment in the Serengeti ecosystem. Besides the importance of water availability and soil fertility, they also found that grazing can actually increase vegetative growth (primary production).

To explore this influence, McNaughton's team fenced off certain areas of the western Serengeti. During their seasonal migration, wildebeest flooded the study site and grazed intensely for four days, consuming roughly 85 percent of plant biomass. In the month that followed, biomass in the inaccessible areas decreased, whereas vegetative growth in the grazed areas increased. As it turns out, grazing increases the growth rate of many grass species. The end result is called "compensatory growth." McNaughton found that this effect is highest at intermediate grazing intensities. Light grazing does not yet produce this effect, whereas very heavy grazing reduces the plant's capacity to recover. In sum, McNaughton and colleagues concluded that large grazing mammals have a substantial influence on the rate of primary production, that they "interact with their habitats in complex and powerful patterns influencing ecosystems for long periods."

Fig. 3.39 Wildebeest migration in Serengeti National Park, Tanzania.
(Daniel Rosengren. CC BY-SA 4.0 International)

To give the students a sense for the immensity of this influence, we remind them that roaming the Serengeti are approximately 1.4 million wildebeest, 600,000 Thompson gazelle, 200,000 zebra, 50,000 buffalo, 60,000 topi, and large numbers of some 20 additional species of grazing mammals.

Together they consume roughly two-thirds of the yearly vegetative growth in the Serengeti. Importantly, the foraging patterns of different grazers are also differentiated. As was mentioned above, in many instances zebras are the first to come and they eat the drier and rougher grasses. The wildebeest follow—strewn out in herds of up to 10,000—and nibble the grass to below ankle height. The really short grass is then grazed by warthogs and gazelles, while the leftovers are nibbled away by Egyptian geese.

There are countless ways that one can explore this universe of relationships and codependency with the students, and, as said above, it helps keep the lessons alive if the teacher continues to explore new avenues and varies the specifics from time to time. After several such examples that show the nature of ecological mutualism very concretely, it can be important to step back and take a more philosophical look at how organisms are commonly viewed with an everyday mindset.

Wolfgang Schad formulates this issue in the following way:

When we study mammalian habitats more closely, we encounter deeply entrenched conceptual obstacles, for landscape is often viewed as a kind of container that feature a set of characteristics that the animal has to learn to live in, somehow adapting secondarily to the prevailing circumstances in order to survive. That this relationship is not monocausal, but interactive, is illustrated by the simple observation that it is not only the habitat that influences the living beings within it, but that these living beings also affect their habitat. Trees affect the soil and climate, and grazers affect their pasture. The fixed "niche concept" considers only the effects of the habitat on its organisms. In real life there is always interdependency.... When we separate parts of the world from their context in order to investigate them, we of course become more familiar with them. But when we then seek to understand their relationship with their environment, it becomes evident that all living beings find themselves in a symbiotic relationship with each other and with their surrounds. The "biosis" of each organism becomes a "symbiosis" when seen in its relationship with other organisms and with the environment, so that finally the distinction between the two concepts ceases to exist. (2020, pp. 695–696)(Gichoki et al. 1996; McNaughton 1985; Molles 2013; Schad 2020).

Mutualism

As an aid in transitioning to other ecosystems, a more general picture of mutualism can help the students see how this form of interaction between organisms is universal and by no means limited to the wilds of Africa. One student favorite is a creature that has a heart rate of around 1,250 beats per minute and the highest energy output per weight-unit of all warm-blooded vertebrates. Put in human terms, their energy output is roughly ten times that of a human being running at nine miles per hour. What's more, these amazing creatures can not only fly forward, but also backward, and even—for a short time—upside down. They are small birds, with most species measuring 3–5 inches in length. They hover in midair with wing-flapping rates that can reach 80 beats per second in the smaller species. By now someone in the class will usually be able to name the animals in question: hummingbirds.

Based on what has just been described, the immediate question becomes: With such an active metabolism, how in the world do these creatures keep themselves going? First of all, their principal food is sugar. They consume half their body weight in sugar daily, which they eat (drink) constantly, refueling five to eight times every hour. But how do they manage this? Well, they have a special group of natural companions: nectar-rich flowers.

Fig. 3.40 Purple-throated Carib Hummingbird [*Eulampis jugularis*].
(Charles J. Sharp. CC BY-SA 3.0 Unported)

To help meet their high-level energy needs, hummingbirds drink nectar, the sweet liquid inside certain flowers. Like bees, they are able to assess the amount of sugar in the nectar they drink. They typically reject flower types that produce nectar that is less than 10% sugar and prefer those whose sugar content is higher. Hummingbirds drink by rapidly lapping nectar with their long tongues. Their tongues have tubes running down their lengths that help them drink more efficiently. In the end, humming birds are only able to consume the amount of nectar they required because many of their morphological traits, such as bill length and curvature, are correlated with the length, curvature, and volume of the corolla of the plants they feed on. The closer the morphological matches, the more efficiently and rapidly the hummingbirds are able to draw nectar. As they hover and insert their bills into flowers that provide them with nourishing nectar, they also return the favor: They brush their heads against the flower's anthers. In their further search for nectar, they will carry this pollen to the stigma of other flowers, pollinating them and thus making it possible for those plants to produce fruit and seeds for a next generation.

Next, we can take the students below ground. In addition to their above-ground symbiosis with the hummingbird, such plants will also have a partnership under the surface that many of the students will have learned about in the context of a farm visit in the 9th grade (see above): the intimate and mutually beneficial symbiotic association between the roots of these plants and fungi. As described earlier, the hyphae of the mycorrhizal fungi extend out from the root system and dramatically increase the total surface area for absorption of water, as well as facilitating the transfer of various mineral nutrients from the soil into the roots. On the other side of the coin, the fungi element in this coexistence receives carbohydrates and vitamins from the plant that are essential for its growth.

Returning above ground, we picture how a deer wanders over to the plant just visited by the hummingbird and browses on it, chewing and swallowing what it takes in. Upon arriving in the deer's stomach, we find—yet again—mutualism at work. By itself a deer is unable to digest much of the plant material it eats. But when we investigate a deer's digestive system, we are met by a variety of microorganisms that make the digestion process possible. Bacteria and protozoa account for 40% to 60% of the microbial mass, with bacteria doing most of the digesting, while the protozoa eat the plant material and degrade the major plant parts. Protozoa also help regulate the gut bacteria population by grazing on them. Along with bacteria and protozoa are fungi, which are important digesters of lignin and hemicellulose. Archaea serve to reduce gaseous buildup by converting methane to carbon dioxide. Finally, there are the viruses, which aren't involved in the digestion of plant matter, but to help keep the other microbes in check. (All of this should come as no surprise to the students due to the extensive discussions about the human microbiome that took place during their 10th grade biology block.)

We already know, of course, what will probably happen with the deer's droppings that fall upon the soil....

More examples can follow, but at some point the wealth of details becomes indigestible for the students and it is worthwhile to move again to the bigger picture.

The significance of such relationships is summed up very clearly by University of Arizona ecologist Judith Bronstein (2015):

> *Mutualisms occur in every aquatic and terrestrial habitat; indeed, ecologists now believe that almost every species on earth is involved directly or indirectly in one or more of these interactions. Mutualisms are crucial to the reproduction and survival of many plants and animals, and to nutrient cycles in ecosystems. Moreover, the ecosystem services mutualists provide (e.g., seed dispersal, pollination, and carbon, nitrogen, and phosphorus cycles resulting from plant-microbe interactions) are leading mutualisms to be increasingly considered conservation priorities (p. 1) .... Mutualism is everywhere. In tropical rainforests, the vast majority of flowering plants depend on mutualistic pollinators and seed dispersers. Deserts are dominated by nitrogen-fixing legumes, lichens dominate tundra habitats, and most northern hardwood forests and grassland plant species depend*

on mycorrhizal fungi to survive and persist. In the ocean, both coral and deep-sea vent communities are rich with mutualisms; coral itself is the product of a mutualistic symbiosis. Organisms from every kingdom have mutualists, and certain kingdoms may in fact have originated as a consequence of symbiotic innovations. (p. 3) (Bronstein 2015; Greenwalt 1990; Maglianesi et al. 2014; Molles 2013; Raven et al. 2012; Raven et al. 2016; Stokes & Stokes 1989; Wildlife Online 2021; Wolf et al. 2021).

North America

After laying the foundation for further explorations through a large scale, multiple biome study such as the one just attempted with Africa, it is easy to move into other ecological contexts and dimensions.

After learning that savannas are the most predominant type of vegetation in the tropics and sub tropics and that, as we saw in Africa, they are located primarily as horizontal (E-W) bands between equatorial rain forests and mid-latitude deserts, the students are surprised to hear that a significant savanna region once ran vertically (N-S) through North America. Prior to the time of European settlement, Midwest oak savannas covered some 50 million acres. In the United States alone it extended over portions of Minnesota, Wisconsin, Illinois, Iowa, Indiana, and Missouri, and down into Nebraska, Oklahoma and Texas. By 1985, however, only 113 sites of really high-quality savanna could still be found in the Midwest, which represent less than 0.02 percent of the presettlement area.

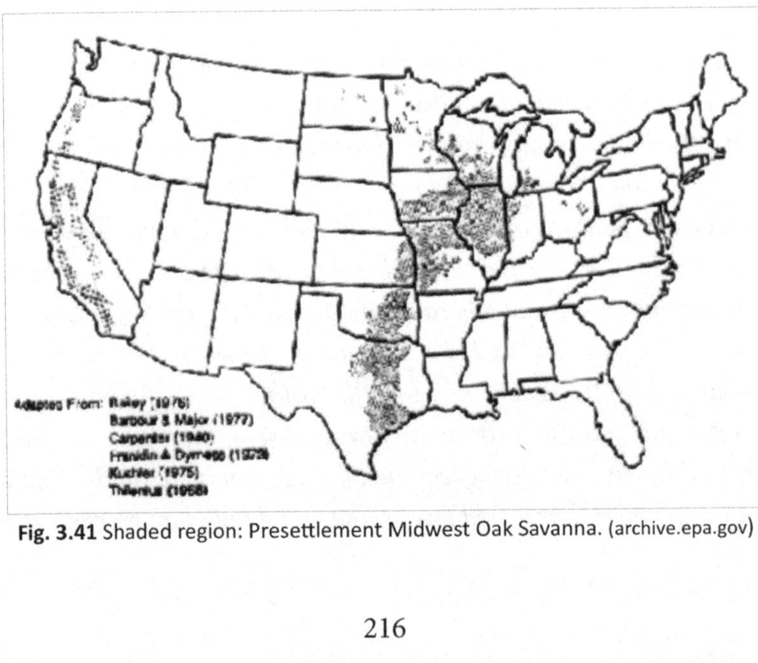

Fig. 3.41 Shaded region: Presettlement Midwest Oak Savanna. (archive.epa.gov)

Based on what they have learned so far about savannas, the students usually find it riddlesome that a savanna could exist at middle latitudes and run north to south, with no rainforest or deserts nearby. If this block is being taught in the US, it is possible to piece together with the student's help a rough picture of what the landscapes to the east and west of the section shaded grey in the map above. Students who live in Illinois, for example, will know that when you drive west toward Colorado, the landscape and vegetation gradually begins to change. As they move across Nebraska on Interstate 80, the trees become fewer and fewer (except for those that line the rivers) until, by the time they have reached 100th meridian (approximately the turnoff to Holdrege, Nebraska—about two-thirds of the way across the state) the open skies and surrounding grasslands give them the distinct impression of having arrived in the West, of having entered a shortgrass prairie landscape. Driving eastward from Chicago, on the other hand, the growing density of wooded areas is soon noticeable and—aside from agricultural influences—does not abate.

The next, obvious question to put before the students is: Why are there such significant changes in the vegetation when one moves from east to west across the middle of the country? Remembering their 10th grade Geophysiology/meteorology block, one of the students will bring up the west to east direction of the prevailing winds at the latitudes we are considering. Looking westward, one finds the Rocky Mountains and the Coastal Ranges. What effect do they have on the prevailing

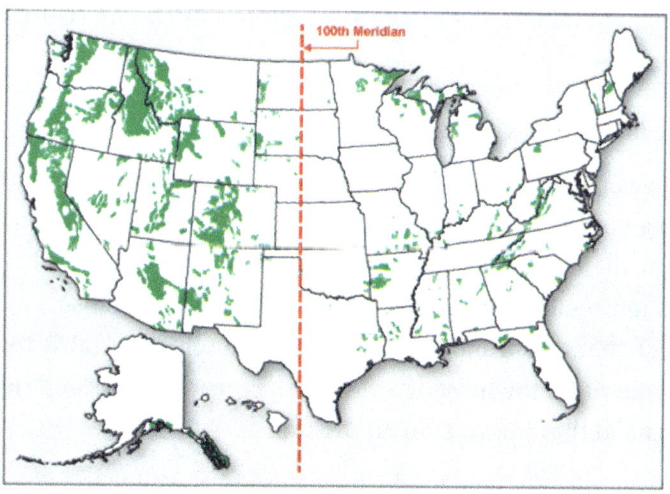

Sources: GAO analysis of U.S. Geological Survey's National Atlas website data. Map Resources (map). | GAO-20-175R

Fig. 3.42 The 100th Meridian.
(Green indicates lands managed by the Forest Service.)

westerly winds, would be the next question to ponder? Further discussion leads to the realization that they cause the winds that rise over them to cool off and release the moisture they carry. This creates a rain-shadow downstream to the east of the mountains. This rain shadow is the primary condition leading to the shortgrass prairie that is predominant from the eastern edge of the mountains to about the 100th meridian, which coincides roughly with the 50-cm rainfall line, the 600-meter elevation line, and the boundary between the shortgrass and the tallgrass prairie.

Whereas the western states are known for their dry air and blue skies, humidity and grey skies become increasingly common as we move eastward. Air flow sweeping up from the Gulf of Mexico becomes a major hydrating factor as we come closer to the forest-dominated east-central regions of the U.S. In short, in the heart of North America we discover a moderate variation on the theme of polarity and transitioning middle that we learned about when discussing Africa. The axis has now shifted from N-S to E-W and the polarity is much less extreme. We find a milder form of the vegetation-sparse desert in the treeless prairie, oaks replacing Acacias and Baobabs in the Midwestern savanna, and temperate forest stands in place of the tropical rainforest.

After an overview of this kind, it is possible to go into much more detail, but, as time is not limitless, choices as to where one goes into more detail must be made. Those decisions are often based on what part of the country we are teaching in and what other regions the students may be familiar with. Here, I will just touch on a few aspects of each of the three ecosystems just mentioned, and will do so primarily in regard to their characteristics prior to European settlement. (Curtis 1959; Nuzzo 1986; Oak Savannas 2021)

North American Tallgrass Prairie

Early travelers over North America's unfenced, unbroken prairies compared the journey to a voyage upon the ocean. It was not only the windblown waves that moved through the tall grasses, or the largeness of the sky in all directions that awakened this impression, it was also the tracklessness, the absence of landmarks, the way in which the edge of a forest resembled a shoreline, a prairie town resembled a seaport, a solitary tree the mast of a vessel. The great stillness and the sensation of solitude, intensified that impression all the more.

In that primeval grassland world, fire was a given. When the ground was damp, or—what was rarely the case—the air still, fires moved idly across the landscape. When the wind was strong, it drove the fires in a continuous wall of a mile or more over land baked dry by sun and wind, incinerating everything in its path. Following a severe fire, the entire landscape was left ashen gray and dreary. But with the arrival of spring, fresh, vigorous growth burst forth from the unharmed roots and rhizomes of prairie plants now freed from overlying layers of mulch and enriched by ashes. In the prairie tracts that remain today, such revitalizing fire events and the unfolding of plant life continues on as it has for thousands of years. Even before the first grass-tips turn green, for example, the fat buds of the pasque flower push forth and unfold their large purple blooms.

Within the greening grass there follows a procession of flowers such as: pinkish shooting star, white anemones, and yellow puccoons. By June spiderwort and golden alexanders have also appeared.

Full summer brings gay feathers and blazing stars, as well as composites such as cone flowers, blackeyed susans, and compass plants. Legumes also appear in great variety: vetches, wild clovers, prairie turnips, and more. As summer moves to

Fig. 3.43 Pasque Flower [*Anemone patens*]. (Funkervogt. CC BY-SA 3.0 Unported)

Fig. 3.44 Shooting Star [*Dodecatheon meadia*]. (H. Zell. Unported GNU Free Documentation License)

Fig. 3.45 Gay Feathers [*liatris punctata*]. (Matt Lavin. CC BY-SA 2.0 Generic)

Fig. 3.46 Compass Plant [*Silphium laciniatum*]. (Raymond M. Coveney. CC BY-SA 3.0 Unported)

fall, asters and various goldenrods come to the fore. By now, the grasses are ripening their seed heads on stalks that reach up above all other plants on the prairie. They, too, take on shimmering color in the evening sunlight of late summer.

As you move away from the mountains, the rain shadow mentioned above becomes less severe, evapotranspiration becomes less, and grasses grow progressively longer. Despite its vastness, this prairie is dominated by only a handful of grasses: big bluestem, little bluestem, Indian grass, and switch grass.

Of these, big bluestem—which once grew taller than a horse over large areas—is the most dominant. On the prairie, it can be found everywhere and commonly accounts for three-quarters or more of the green cover of the land. Contributing to its dominance are its rapid growth, height (3.5–10 feet), shade tolerance, dense sod-forming habits, and its deep, fibrous root system—which branches profusely and commonly reaches depths of six to seven feet.

Fig. 3.47 Big Bluestem [*Andropogon gerardii*]. (Ammodramus. Public domain)

One striking characteristic of prairie plants is their ability to coexist tightly intermingled in moisture-deficient soil. The dense network of grass roots and rhizomes permeates the top few inches of the ground and is further intertwined with the roots and stems of many other plants, such as sunflowers (*Helianthus* sp.), goldenrods (*Solidago* sp.), mints (*Blephilia* sp.) and the like. The grasses permeate the soil most densely, producing on average, three to four times more roots by weight than they do stems and leaves. This results in a root-to-shoot ratio ten times higher than that of a forest. The students are amazed to hear that all the root fibers produced by a small patch (1.5 meters by 1.5 meters) of Nebraska's tallgrass prairie—if laid end to end—would extend for 150 miles. By wriggling into every nook and cranny of the soil, these slender root fibers hold the soil together. As the roots interweave from plant to plant, the soil becomes tightly bound-up into a thick fibrous mat. The result is the famous "prairie sod" that the settlers used to build the first homes in these treeless lands.

For thousands of years, this root fabric has provided the raw material for decomposers—fungi, nematodes, springtails and the like—that begin the process that leads to thick layers of dark, crumbly, humus-rich earth for which the prairie is known. However, the grasses alone could not have created this "prairie earth," which is some of the richest soil known. The nonliving substrate—the parent material—has also played an important role in this. The soils to the north and east of the Missouri, for example, have been built up from glacial drift; others have been blanketed with thick drifts of loess. Loess is probably the most prevalent parent material in the Midwestern prairies, followed by glacial drift. In both cases, this glacial material has provided the soil of this region with a rich supply of fertility-enhancing plant nutrients. In yet other places, like the Sand Hills of Nebraska, the Ice Age left behind a legacy of sand that has led to distinctive communities of grasses and forbs in those locations.

As indicated earlier, soils on the plains are also strongly influenced by moisture availability: The more moisture, the more roots a plant can produce and the deeper and darker the topsoil that accumulates. It thus follows, that not only precipitation increases as we move east from the Rocky Mountains, but also soil fertility. What's more, warmth also plays a pivotal role in soil development. Warm temperatures in the south increase the speed of biological and chemical processes in the earth, thereby promoting soil formation. Where soil temperatures are lower, such as in the northern plains, soil organisms are less active and the rate of decomposition slower.

In addition to large-scale climatic patterns, the patchwork quilt that makes up the grasslands is also shaped by topography and the microclimates that this brings with it. For example, the orientation of a slope face (its "aspect," or direction of exposure) influences soil temperature. South-facing slopes receive direct sunlight in winter, which melts snow and warms the surface, whereas north-facing slopes are shaded until spring. On the other hand, south-facing slopes are more likely to dry out due to their extended exposure to the sun. Conditions are also different on a hilltop than they are in the lowlands, which reproduces on a small scale the moisture-gradient that exists from west to east. Thus a sea of tallgrass prairie can be broken up by islands of mixed or short grasses that grow on arid slopes or uplands.

The tallgrass prairie also supports a large population of animals. Beneath the surface and in the dense web of stems and leaf-bases live a multitude of mice, ground squirrels, and gophers. These and other burrowing animals, such as moles and badgers, perform the important function of turning over the prairie soil. These smaller creatures are hunted by fox, coyote, and—in times past—wolves. Of the birds, the now rare prairie chicken—a large, ground-dwelling seed eater, whose elaborate courtship behavior takes place on the same specific courting grounds year after year—is uniquely characteristic of the prairie. Hawks soaring in a prairie sky can be seen for miles around, while bobolinks, meadow larks, and horned lark draw our attention by singing as they fly. Insects populate the Midwestern prairie in great number and variety: Grasshoppers, beetles, crickets, leafhoppers, spittlebugs, and a host of other crawlers, runners, leapers and fliers are active in the grasslands. Ants, in particular, are an important part of life on the prairie. Around 30 different ant species live in most prairies. They nest in the soil, creating a complex of tunnels that often extend to a depth of three to six feet. The particles of clay they carry up from the depths help replenish the upper layers of soil. One estimate suggests that the upper two feet of prairie soil is turned over every hundred years by the unending activity of ants and other creatures.

Prior to the early 1800s, every spring for millennia countless bison (*Bison bison*) poured out of their winter range and massed into vast herds that flowed across the great plains in search of green pastures. Each year they charted new routes, moving east to track recent rainstorms, south for the fresh green-up of recently burned grasslands, north to follow the spring.

Fig. 3.48 Bison grazing. (Lee Coursey. CC BY-SA 2.0 Generic

The plain's grasses were fitting partners for huge bison herds. In contrast to shrubs and forbs, whose meristems develop on branches and growing shoots, grass meristems are at, or even slightly below, ground level. This means that if a grass is eaten, it is able to send up fresh leaves immediately, whereas the development of a forb or shrub may be arrested until the next spring. Moreover, many defoliated grasses react by shifting resources from their roots to new shoots, which leads to a dramatic increase in photosynthesis and exuberant growth that goes beyond the mere replacement of tissues that have been removed. Just as fire is now recognized as a vital factor in tallgrass prairie management, the need to reintroduce large ungulate herbivores to these grasslands has also become evident. Similar to fire, bison-grazing reduces standing dead biomass. Moreover, the way in which bison graze—both over time and spatially—has a significant impact on biodiversity. Such grazing-induced heterogeneity stands in strong contrast to the spatial homogeneity that follows from fire in an ungrazed landscape. Because they prefer grasses, bison leave most forbs untouched, which results in much greater plant diversity in the areas they have passed through. More wild flowers mean more insects and other invertebrates. Students who do not come from former bison country are often surprised and fascinated to hear of another way that Bison further plant diversity: by wallowing.

Bison love to roll around and rub their bodies on the ground, creating bare, dusty patches that become concave depressions in the earth through repeated use.

Fig. 3.49 A bison wallowing. (Brocken Inaglory. CC BY-SA 3.0 Unported)

The prairie where they live becomes pocketed with many such wallows. Due to the compacted soil and the bowl-like shape, wallows hold rainwater better than their surroundings, creating a unique microenvironment for a greater variety of plant species, as well as watering holes in times of drought. With its furry coat, the bison is also a first-rate seed spreader, not to mention the seeds it distributes through its digestive tract as it moves around.

The bison's digestive system is—as the students should be expecting—an ecosystem of its own (a microbiome!) populated by countless microorganisms that are engaged in digesting the huge quantities of grass that the bison consume daily. These microbes and undigested plant matter are passed on to the prairie soil in huge quantities as feces (buffalo chips) and urine. Each bison gives 10 to 12 quarts of dung and gallons of urine to the soil each day. (We can let the students multiply those quantities by 50 million to give them an idea of the huge effect this must have had on the prairies where the bison once roamed. Not only is this an important source of nitrogen, phosphorous, calcium, etc. for microbes, plants, and other animals, but bison dung also hosts a rich community of insects, with as many as 300 species of insects living in one bison patty. This also makes the patties a delightful resource for bats, box turtles and birds. As the students should be expecting, one special resident of the patty that has particular significance for the soil is the dung beetle. These tunnelers and rollers are specialized on the dung of bison and bury much of the fresh manure, thereby activating the decomposing microbes that make the contents of the manure more accessible to plants.

Much more could be said about the life and significance of the bison herds in presettlement times, but one aspect that touches the students deeply is the role that the bison played in the life of different Native American tribes. They were central for the Plains Indians not only as a food source, but in countless other ways—both practical and spiritual. Rather than merely providing a short sketch of these important aspects here, I refer the reader to Tom McHugh's fascinating depiction of this topic in his book *The Time of the Buffalo* (1979), and the chapter on the American bison in Craig Holdrege's book *Seeing the Animal Whole* (2021). These resources can provide the teacher with a very living and pedagogically effective picture of this deep human–animal relationship that was so central in the history of the North American continent.

What remains of the tallgrass prairie in the central United States is among North America's most endangered ecosystems. It occupies a zone of tension between more arid grasslands to the west and more mesic forest to the east, which makes it particularly sensitive to shifts in land management, climate and fire regimes. Most of this ecosystem was lost when the land was converted to row-crop farming. The greatest threat to remaining tracts of mesic grasslands is the expansion of woody vegetation into these areas, which in the past was limited by widespread fire. (Anderson 2006; Axelrod 1985; Barlow 2003; Briggs et al. 2005; Chadwick 1995; Collins et al. 1998; Heisler et al. 2004; Holdrege 2021; Kirt 2000; Knapp 1999; Madson 2004; Manning 1995; McHugh 1979; National Park Service 2016; Pritchett & Fisher 1987; Reichman 1987; Savage 2004; Thomson 1976; Thomson 1977a; Winckler 2004)

Eastern Forests

Moving to the presettlelment forests of the Eastern U.S. we enter a very different world. Whereas early travelers across the unbroken prairies felt they were crossing an ocean on its surface, those who entered the forest primeval in the East also spoke of an "ocean of woods," but from their descriptions they were not travelling on the surface, but under water! They spoke of "one immense uninterrupted forest" where "a squirrel might have leaped from bough to bough for a thousand miles and never have seen a flicker of sunshine on the ground." To early European visitors looking up from the ground, the tops of the trees were "so close to one another from many miles together, that there was no seeing which way the clouds drive, nor which way the wind sets: And it seemed almost as if the sun had never shone on the ground, since the creation." (Whitney 1996, pp. 53–54) One early Massachusetts forester, George Emerson, observed that there were more trees in the state of Massachusetts than in any country in Europe—trees of tremendous diversity and grandeur.

Another visitor to those woods observed how "one of the conditions most remarkable in a primitive forest is the universal dampness of the ground… the crowded assemblage of trees and their undergrowth, causing great difficulty in traversing it. Innumerable straggling vines, many of them covered with thorns, like the green-brier, intercept our way. Immense trunks of trees, prostrated by hurricanes, lie in our path, and beds of moss of extreme thickness cover a great part of the surface, saturated with moisture. The trees are also covered with mosses, generated by the shade and dampness (Whitney 1996, p. 57)."

In the forests of the northeast, the average canopy height was around 80 feet, with an emergent layer of white pine reaching up to 150ft. In some areas the white pine would even reach 250 feet in height. The trees were typically from four- to six-feet in diameter (six feet = a circumference of 38 feet) and often stood only four- to six-feet apart from one another, with no branches close the ground. Further west in the dense forests of the Ohio and Wabash River Valleys, visitors would sometimes encounter sycamore trees with a circumference of 50 to 60 feet.

The European visitors to New England were left dumbfounded by the incredible abundance of life—both plant and animal—that they encountered there. Along the coast they were astonished by the spring spawning runs of sturgeon, smelt and other ocean fish, which were migrating to fresh water to deposit their eggs. The fish pressed up the shallow streams in such multitudes that some observers fancied they might be able to cross a steam with dry feet by walking on their backs. One individual spoke of walking across a beach for a quarter mile knee-deep in stranded herring. The movement of the tides would bring thousands of lobsters into the shallow waters offshore, not to mention the countless mussels and other shellfish they exposed. The abundance of birds was no less amazing. Waterfowl hunters were able to kill a hundred geese in a week, a dozen wild turkeys in a half-day. The semiannual flights of the passenger pigeons were beyond belief. Without beginning or end, millions of millions passed overhead, so densely packed that the sun could not be seen. Flocks of doves, containing many thousands each, were also so dense that they too would obscure the light. Mammals, too, were present in large numbers: bear, deer (in the spring one could see as many as a hundred deer in the space of a mile), moose, wolves, foxes, wild cats, martins, otters. Although mice were prevalent, there were no rats.

To describe the richness of nature in the Eastern forests in this way without mentioning the people who had been living there for thousands of years would mean leaving out a central influence on the whole ecosystem. These is much more to tell than I will describe in a short overview, and for students who live in this part of the world, it is an important and highly interesting subject to explore in more detail. Suffice it to say that the Native Americans in New England had an intimate understanding of the ecological relationships that formed their environments there. Their villages were not fixed entities, but changed in size and location on a seasonal basis. Communities would breakup and reassemble based on the habitats available to them within seasonal cycles. All aspects of the Native American life were built on mobility, with houses designed to be taken apart and moved within hours. Tools and other properties were either light and easy to carry, or left behind and remade when needed. Those in the more northerly regions lived entirely as hunter-gatherers, further south they would raise crops as part of their annual subsistence cycles.

In the northern forests of New England the Native American populations were low and had relatively little impact on the ecosystems they lived in. In Southern New England they not only engaged in hunting and gathering, but also raised crops. This put them in a very different relationship to their environment. For one, it led to much greater population densities. Whereas in the northern regions you had approximately 40 hunter-gatherers per 100 square miles, in the more southerly crop-raising areas, roughly seven times as many people could be sustained in an area of similar size. As a result, although the native peoples occupied similar amounts of land in the north and south of New England, over 80 percent of the population was in the southern crop-raising area.

Crops were planted between May and late June. Once planted and weeded they needed little attention for two to three months until, prior to harvesting, the ripening corn required protection from marauding birds. During those months families would often move their individual wigwams to other sites where the women planted and gathered, while the men fanned out for extended hunting and fishing trips. In Autumn the corn was harvested, together with the gathering of chestnuts, acorns and other wild plants. Many hundreds of people would also gather at that time for extensive festivals, during which they would consume much of the surplus food.

But once the harvest celebrations were over, the bulk of corn and beans would be stored, the households struck, and everyone would move to campsites for the fall hunt. From October to December villages broke into small bands in search of animals like deer and bear, which at that time of year were at their fattest. The skins of these and other furbearing animals would provide the village with clothing for the coming year. After a winter of hunting and trapping, small groups in some areas would gather in the spring to tap maple trees and boil the sap down to hard sugar.

By clearing land for planting, the Native Americans of southern New England took an important step in reshaping the ecosystems they lived in. They would return to the same field site for as long as possible—usually eight to ten years—until the soil gradually lost its fertility and necessitated creating a new field. The yearly reoccupations of fixed planting and village sites meant that the area around the camp and field would experience intensive food gathering, garbage accumulation and the consumption of firewood. But villagers' influence on the environment went beyond this. They also engaged in the burning of extensive sections of surrounding forest once or twice a year. This consumed all the underwood that would otherwise overgrow the area, making it unpassable and hunting difficult. It resulted in a park-like woodland of large, widely-spaced trees, with few shrubs and much herbage and grass. Due to their yearly repetition, the fires had, for the most part, only small non-woody plants to consume and thus burnt at relatively low temperatures, moved rapidly, and soon extinguished themselves. These were not forest fires as we know them today, but ground fires that did not usually involve large trees.

Such underwood burning had additional, more subtle ecological effects. Because they increased the rate that forest nutrients were recycled into the soil, grasses and other plants tended to grow more luxuriantly after a fire. These fires created favorable conditions for many gatherable foods such as blackberries, raspberries and strawberries. Due to the elimination of smaller trees and the thinning of the forest canopy through burning more light was able to reach the forest floor. Even grasses like the little bluestem—nonexistent in dense forests—became abundant in these areas. The forest soil also became warmer and drier, which discouraged some tree species and furthered others. In sum, selective burning by the Native Americans led to a mosaic quality in New England ecosystems and what is known as the "edge effect," where boundary areas between woods and open areas create an ideal habitat for many wildlife species. By increasing the herbivorous food supply in this way, these edge areas promoted a significant increase in the species that so impressed the English colonists: deer, elk, turkey, beaver, hare, quail, porcupine, and so on. With the increase in these species came a corresponding one in their carnivorous counterparts: wolves, foxes, lynxes, eagles, hawks. In short, the Native Americans were not just the recipients of nature's abundance, they helped bring it about through controlled burning and other activities. Not only that, they kept their demands on these ecosystems to a minimum by moving their settlements from habitat to habitat: Their way of life corresponded to the ecological diversity they helped create. (Cronon 1983; Johnson 2002; Sokolow 2003; Thomson 1977b; Weitzel 2021; Whitney 1996; Zimmerman & Molyneaux 1996)

Midwestern Oak Savannas

After covering the midwestern/western prairies of the United States and its eastern forests as a moderate variation on the desert-rainforest polarity that we found in Africa, we come to the mediating ecosystem between the poles: the midwestern oak savannas. To describe them as a modest manifestation of what we find in the magnificent savannas of Africa may seem somewhat of a stretch. It is only possible to do so if we emphasize to the students—as mentioned earlier—that only 0.02 percent of the presettlement oak savannas still exist today. What once covered 50,000,000 acres has been reduced to mostly degraded plots of less than 100 acres hidden away in various locations across the Midwest. Due to their rarity today, these savannas have been classified as "globally imperiled."

Where did they go? Because of their openness, savannas appealed strongly to the early settlers who were moving westward out of the dense eastern forests and were thus among the first areas to be settled in the Midwest. Whereas the open areas were suitable for agriculture, for buildings the oak groves along the edges provided shade from the intense summer heat. The grasses of the open areas were plowed under and many of the oaks cut down or their roots damaged by digging and travel over them. The herbaceous layer was quickly gone due to trampling and mowing of tall "weeds," or through conversion to pasture. The original bluestem grasses, as well as other prairie and savanna plants, disappeared rapidly under grazing and were replaced by invasive species of European origin. And, of course, all fires were suppressed. Within a few years of settlement, the specific savanna character of most areas was unalterably ruined.

As in the eastern forests, the oak savannas of the Midwest were influenced by the Native Americans who lived in the region. In the Chicago area, for example, the Potawatomi tilled the soil in places and helped, as hunters, keep the deer population in check. Most importantly, they used the cleansing effect of fire to prevent the savanna areas from filling in with brush, shrubs and fire intolerant trees, which otherwise would create woods that were poor food producers and difficult to hunt in.

Fig. 3.50 Prescribed burn — Oak Savannah in Iowa.
(Rbreidbrown. CC BY-SA 4.0 International)

Where a regular fire cycle was maintained, bur oak (*Quercus macrocarpa*) was the predominant savanna tree species. There are several reasons for this. For one, when the acorn of a bur oak sprouts, it already sends down a tap root about nine inches long before the first leaves unfurl. In the first year that root easily reaches a length of three feet. At the age of three, the roots can reach a depth of eight-and-one-half feet, with a spread of more than five feet. Once established, they develop heavy trunks and crooked, gnarly branches armored with rough bark so thick that the twigs become as fat as a little finger. This thick, cork-like bark makes the bur oak tolerant of fires that most trees cannot withstand. What is more, Oak leaves, together with grasses, are a primary fuel of oak savanna fires. Oak leaves burn much more readily than the leaves and litter of other hardwoods. There are several reasons for this. Oak leaves are much thicker than other hardwood leaves and contain tannins, giving them greater resistance to decomposition and longer life spans in the leaf litter. They also tend to be drier (more xerophytic), which makes them more flammable. What's more, oak leaves curl up more than most leaves when they hit the ground, which raises the fire up off the surface and helps it spread more effectively.

After determining the suitability of bur oaks for an environment frequently visited by fire, the important question becomes: What role do these oaks play in creating an ecosystem that is not just a tree-studded prairie, but one that is unique in its own right? For one, oak savannas are very heterogeneous in the density of trees from place to place. This results in significant variations in the availability of light across a savanna. It creates a mosaic of shaded and sunlit areas that also change

Fig. 3.51 Freestanding Bur Oak.
(Justin Meissen. CC BY-SA 2.0 Generic)

Fig. 3.52 Minnesota Oak Savanna. (Paul Nelson: CC BY-SA 3.0 Unported)

during the course of a day with the path of the sun. In addition to the sun's direct effect on plant growth, the prevalence or lack of sunshine affects soil temperature and moisture levels. Since plants vary in their temperature requirements and in their root-system-related drought tolerance, the extent of canopy openness or closure has a major effect on the plants that will thrive there.

Very open savannas will have more prairie plants, more closed ones will tend toward woodland species. Other species—known as savanna specialists—thrive in conditions intermediate between very open and mostly closed canopy. Because of its wide variations in canopy, the number of species found in a well-restored savanna will usually be much higher than in a well-restored prairie or forest.

One reason for this is that the north and south sides of savanna trees provide quite different environments for the plants growing there. In the less shady, warmer south side, grasses and forbs tend to thrive that are typically found in the prairie, whereas the shadier north side is often home to woodland species. In some instances, large freestanding oaks have been known to have over 20 plant species growing in the space influenced by their presence.

Related to the above is the phenomenon that oak savannas generally develop on south- or southwest-facing slopes where they receive more sunlight and are drier. Woodlands, on the other hand, tend to develop in moister areas, on north- or northeast-facing slopes, or in lowlands. In contrast to freestanding oaks with well-developed, wide-reaching lower limbs, a healthy woodland consists of tall straight trees without significant lower branches, but whose top canopies spread out high above, where they gain more access to light.

Fig. 3.53 Oak savanna habitat supports a wide variety of wildlife. With prairie-like and woodland-like areas, it is home to red-headed woodpeckers, skinks, badgers, deer, wild turkey, pollinators, and more! (USFWSmidwest. CC BY-SA 2.0 Generic)

Another factor contributing to the flower diversity in oak savannas is the presence of both "short-day" and "long-day" forbs. In many plants the flowering process is influenced by the length of the day rather than the amount of sunlight they receive. Most short-day plants—such as goldenrods, asters and ragweeds—bloom mainly in September/October when the days have grown short.

Whereas long day plants—such as cone flowers and black-eyed Susans—grow vegetatively throughout the spring but do not begin to flower till midsummer.

Fig. 3.54 Zig-zag Goldenrod [*Solidago flexicaulis*]: a woodland species often found growing on the shady and cooler north side of a freestanding bur oak. (R.A. Nonenmacher. CC BY-SA 4.0)

Fig. 3.55 Gray Goldenrod [*Solidago nemoralis*]: a dry prairie species often found growing on the sunny and warmer south side of a freestanding bur oak. (Daniel McClosky. CC BY-SA 4.0)

Fig. 3.56 New England Aster [*Aster nova-angliae*] is a short-day plant that begins to flower in late August and is at its peak in September. (Haeferl. CC BY-SA 3.0 Unported)

Fig. 3.57 Black-eyed Susan [*Rudbeckia hirta*]. A typical long-day plant, which is in full flower during midsummer. (Pandries. CC BY-SA 4.0)

Once returned to their former glory, midwestern oak savannas erupt in color and biological diversity. One restored savanna in the Chicago area, Vestal Grove, blazes with color from May through October. Early in the growing year, plants that mature low to the ground dominate. Progressively taller types take over as spring turns to summer and then summer to fall. From March to June short-stalked wildflowers display flowery heads of predominantly yellow and white, whereas midsummer's taller flowering plants bring not only a celebration of whites and yellows, but also of blue, violet, lavender, pink, and red with bright green as their background. Pale purple, lilac, and dull pink flowers become more prominent as we move from August through the autumn. Innumerable additional flowers unfold as a sea of tall prairie grasses, with big bluestem—characterized by its turkey-foot-like seed head—and shiny golden-brown Indian grass being the most prominent.

Oak savannas also provide outstanding conditions for a variety of wildlife. One reason for this is that of all forest trees, oaks are the most important provider of animal food. Countless oak acorns fall to the ground and produce fodder for wildlife. They are called "mast"—from the old English term *maest*—which means the nuts of forest trees that have accumulated on the ground). A mast year is when the oaks produce an extra-large crop of acorns. Bur oaks tend to follow a two-year cycle in this regard. The reproductive cycle of some animals—including deer, bear, and squirrels—depend on mast. Although long-lived animals like deer do not experience sharp population fluctuations, the populations of shorter-lived creatures like rodents fluctuate with mast cycles. Oak mast is also central in the diet of other savanna dwellers, ranging from wood turtles and ruffled grouse to racoons and rabbits.

Not surprisingly, birds are also prominent savanna residents, with wild turkeys—the largest game bird in North America—playing a significant role. Their preferred habitat for roosting is a large tree with horizontal limbs, which open-grown oaks are happy to provide. Although omnivorous, they prefer acorns to all other foods.

Fig. 3.58 Male Wild Turkey strutting. (Frank Schulenburg. CC BY-SA 4.0)

Blue jays, in particular, play a key role in oak savanna ecosystems. Amazingly, blue jays cache between 3,000–5,000 acorns each year. This means that if each jay "forgets" only 5 percent of its crop, an oak savanna will still have thousands of new oaks germinating each year. Important in this context is the blue jay's ability to distinguish fertile acorns from others with 88 percent accuracy. (Infertile acorns are often infested with fungus, rust or acorn weevil.) One study showed that blue jays cached 55 percent of the acorns in a given area and ate another 20 percent along the way. One reason blue jays can move such huge quantities is because they are

Fig. 3.59 Blue Jay. (Rob Hanson. CC BY-SA 2.0 Generic)

able to transport numerous acorns at a time, carrying two to three in their gular pouch, one or two in their mouths, and one more on the tip of the beak. Often these acorns will be carried over 2.5 miles from the source tree, thereby supporting tree dispersal. Squirrels, by contrast, usually bury acorns only a few feet from where they find them.

Oak savannas are also a favorite habitat for cavity-nesting birds, who either excavate their own holes, use cavities resulting from decay, or occupy holes created by other species. The trees they use—known as snags—are dead or deteriorating. Most of the birds eat insects, which is an important factor in pest control. One key cavity-nester is the red-headed woodpecker.

Fig. 3.60 Red-headed Woodpecker. (USFWS Midwest Region. CC BY-SA 2.0 Generic)

Red-headed woodpeckers are omnivorous with a very diverse diet. They consume and cache acorns which they supplement with insects, fruits, seeds, eggs, mice, and even snakes and lizards. In the summer, insects–which they are experts at catching in midair—make up a large proportion of their prey. Restoration efforts have very important implications for them in this regard. The clearing of invasive brush and trees opens up the vistas that allow them to detect their prey. Leaving dead trees and snags instead of removing them for firewood also provides the nesting-cavity habitat that they need. Their numbers have declined greatly over the past decades and because oak savannas are ideal for these birds, there is great hope that continued savanna restoration will lead to their revival. Where such initiatives have been successful, red-headed woodpeckers sometimes appear not only as summer visitors, but also overwinter in those savannas.

Other prominent cavity nesting birds that populate midwestern oak savannas include the pileated woodpecker, yellow-bellied sapsucker, hairy woodpecker, downy woodpecker, American kestrel, black capped chickadee, tree swallow, great-crested fly catcher, tufted titmouse, starling, and more, Other interesting birds

Fig. 3.61 Barred Owl.
(Andy Reago & Chrissy McClarren. CC BY-SA 2.0 Generic)

Fig. 3.62 Cooper's Hawk.
(Matt Tillett. CC BY-SA 2.0 Generic)

to pursue would be bobolinks, sandhill cranes, wood turtles, trumpeter swans, northern harriers, dickcissels, Ozark minnows, redfin shiners and raptors like the barred owl, and Cooper's hawk.

Needless to say, there is so much to discover in Oak savannas that a whole course could be developed just pertaining to them. In concluding this topic, it is important to emphasize oak Savannas are not static entities. They are interlaced with forests, prairies and wetlands in many different ways that can shift over time. In the words of William Stevens (1995):

> *The oak savannas of the mid-continent in presettlement times were an infinitely complex array of types, depending on soils, climate, fire history, and stage of succession. But together they constituted a vast landscape stretching down through the country's midsection all the way from the Great Lakes to the Gulf Coast.*

After such an overview of three major ecosystems, one can go into much more detail depending on where one is located. This is most effectively done if it can be combined with field trips into local sites—ideally where restoration efforts are under way. Finding possibilities for the students to engage in such initiatives in the context of service learning is one excellent way to provide a hands-on experience of specific ecosystems and the kind of efforts required to preserve them in the midst of the modern world where economic forces often have the upper hand. (Bray 1960; Curtis 1959; Dey 2021; Harry 2018; Koenig & Knops, 2002; Leach & Givnish1988; McPherson 1997; McShea & Healy 2002; Minnesota Dept. of Natural Resources 2006; National Wild Turkey Federation 2022; Nelson 2017; Oak Savannas 2021; Savannah Oak Foundation 2021; Stevens 1995; Thomson 1977b)

Additional Topics: So Many Possibilities

Equator to Arctic

Although I have taken the African continent and North America as examples for introducing the students to a very fundamental ecological polarity, there are many other exciting possibilities for this as well. If mountains are anywhere nearby, one interesting approach is to explore a mountain from its base below all the way to the peaks high above and to compare what we find there with the way that the vegetative life of the earth changes as one moves from the equator to the arctic. In the latter case, the changes in vegetation are intimately connected with the path of the sun. At the equator the sun rises and sets vertically, with its arc only shifting 23.5 degrees to the north and to the south in the course of a year. Intense sunlight, warmth and moisture are a given. With them, as we have already heard, comes an abundance of life, with gigantic trees rising high above the forest floor, each one like a garden covered by epiphytes that live in their crowns, branches and hollows—and whose flowers are constantly visited by butterflies, birds, and a whole community of other tree dwelling animals.

In the arctic, the sun shines half the year without setting, but always at an angle so low that the growing season lasts less than 50 days, with a mean July temperature of around 40° F. With the power of the sun so diminished, not only does water remain frozen most of the year, but the vegetative world presents a very different picture from further south.

Fig. 3.63 Arctic tundra. (ADialla. CC BY-SA 2.0 Generic)

There are no trees; mosses and lichens are prevalent, and all plants remain close to the ground. Arctic forbs and shrubs are hardened, finely formed and contracted to dwarfness. They often form low mounds or cushions with hemispherical shapes. Luxuriant vegetative growth never develops here, and with a permanently frozen subsoil (permafrost), there are no deep roots. Moving to the southern boundary of the tundra, where isolated trees begin to appear, we find the first indication of vertical unfolding. For example, the stems of many spruce branch out horizontally and directly over the surface, before finally turning up their tips—but only at the very periphery.

Fig. 3.64 Spruce at the southern periphery of the arctic tundra.
(Daniel Chase. CC BY-SA 3.0 Unported)

Arctic willows show a similar growth form. Yet, despite all the contracting and hardening forces that are manifest there, during the very short summer the mats and cushions produce an abundance of colorful, often star-like flowers that shine above the stunted vegetation.

Fig. 3.65 Arctic flowers.
(Penny Higgins. CC BY-SA 2.0 Generic)

A diverse and interesting animal population also makes the arctic tundra its home. It ranges from caribou, lemmings and arctic hares, to wolves, foxes and polar bears—not to mention the migratory birds and insects that can be found there. Some of the animals are active all winter, some hibernate, and others migrate further south.

When we move further south into the boreal forest, it is not surprising to find that certain characteristics reminiscent of the tundra remain. The northernmost forest consists of a single canopy (tree layer) consisting primarily of conifers.

Fig. 3.66 Boreal lake and forest in Northern Quebec.
(Peupleloup. CC BY-SA 3.0 Unported)

These are trees whose leaves have not only contracted to the degree that they are needle-like, but they—as "ever-greens"—have emancipated themselves from the yearly, sun-linked rhythm of shedding and re-growing leaves, and are able to photosynthesize whenever temperatures permit. The branches of these trees do not spread out and open themselves to the light and air around them, but remain in tight whorls around a central, pillar-like trunk. What's more, the seed-bearing cones have hardened to wood. The difference in mood one experiences when walking in a dense forest of spruce compared to one composed of deciduous trees is striking. The former appears dark, earnest and ancient, the latter young and cheerful, by comparison.

Moving yet further south into the temperate regions, the endless, largely monotone, coniferous forest opens up into a great diversity of ecosystems, only a small part of which was discussed in our "prairie—savanna—eastern forests" comparison above. In short, the north–south comparison moving from the artic to the equator, from tundra to tropics—or visa versa—has much to offer. (Breckle 2002; Grohmann 1989; Molles 2013; Odum & Barrett 2005; UCMP 2021; Woodward 2003)

Valley Floor to Mountain Peak

Such comparisons are particularly interesting for the students if they have the opportunity to explore for themselves what a nearby mountain range can reveal when they move—instead of in the north–south axis—in a new dimension, from below to above, from the foot of a mountain all the way to the—often ice-crowned—top.

When I taught in Austria, I had the fantastic opportunity to go with students into the alps for a week each year, where we would begin with the vegetation-rich, mountain valleys below, and gradually work our way upward—day by day—through the changing alpine forest to the windswept, treeless alpine tundra that often ended at the edge of a glacier. We would observe, draw/paint, and in the evening discuss what we had experienced each day. In the words of the botanist C. Schroeter (1989):

> *The way from the plains up into the arctic of the high peaks takes the mountaineer through an instructive compendium of the different zones. A walk on the southern slopes of the Alps is especially exciting: We turn away from the Lago Maggiore (196.5 meters above sea level), the olive trees, cypresses, oranges and pomegranate trees of the a Mediterranean climes and their chestnut groves, and walk in the shade of central European beech woods; then we inhale the aromatic fragrance of the Nordic spruce and the larch forests, enjoy the beauty of the flower-studded alpine meadows above the tree line and find ourselves surrounded by a great number of plants of arctic type, right up to the summit of Pizzo Centrale (3003 meters).*
>
> *Here we stand among plants from Spitzbergen and Greenland, and look down into the southern valleys. So it happens that 2800 meters difference in altitude is equal to a journey from the 40th to the 80th parallel, a distance of about 9000 kilometers, while the distance from Locarno on the Lago Maggiore to the Pizzo Centrale is 48 kilometers, as the crow flies.*

Fig. 3.67 View to Lake Maggiore.
(Zairon. CC BY-SA 4.0 International)

Fig. 3.68 Pizzo Centrale.
(Albinfo. CC BY-SA 3.0 Unported)

As a Colorado native, I have often explored the delightful transitions that exist when moving from the grasslands below to the mountain tops above, where again, one meets a location-specific variation on the kind of polarity that we used as our starting point in the pages above. Mountain ranges throughout North America provide similar opportunities of this kind. As always, there are many interesting possibilities to move into greater detail, to hone in on this or that interesting aspect of the environment without losing the larger context in which it is embedded.

Depending on where one lives, it may make more sense to focus on wetland, riparian and upland habitats, or on a lake or ocean if there is one nearby—comparing and contrasting the various elements in a manner similar to what we have attempted above. And, of course, it should not be forgotten that besides in the Midwest, oak savannas of various flavors can also be found in California, Washington and Oregon, as well as in the Southwestern US and Mexico.

Ecological Succession

After getting to know various ecosystems such as those described above, the students are often interested in learning how ecosystems arise. In earlier blocks they have learned how organisms go through lawful stages of development: how they are neither put together piece by piece, nor do they follow an arbitrary timeline as they gradually unfold. In this course, the students have also learned how ecosystems involve organisms that are interwoven with each other and with other elements of their environment. So the question becomes: How does such a web of relationships come into being?

The process by which a biological community evolves over time is called ecological succession. Ecologists distinguish between two different types of succession—primary and secondary. Primary succession begins in essentially lifeless areas—regions where the soil is incapable of supporting life as a result of factors like retreating glaciers, lava flows, and newly formed sand dunes.

Secondary succession occurs in areas where a community that previously existed has been disturbed, but not all life and nutrients have been removed.

Both forms of succession create a continually changing mix of species within communities. The sequential progression of species during succession is not random. Initially only a small number of species from surrounding habitats (pioneer species) can thrive in the disturbed habitat. As these plant species take hold, they modify the habitat by altering such things as soil nutrients and the amount of shade on the ground. With time, other species that are better suited to this modified habitat succeed the pioneer species. These newer species are gradually replaced, in turn, by still newer ones. A succession of animal species occurs in a similar fashion, and interactions between plants, animals, and environmental conditions shape the pattern and rate of successional change. In some environments, succession will result, lastly, in a state of equilibrium, in a stable climax community.

The first generation of ecologists to study succession were so deeply impressed by the lawful development they observed that they spoke of such plant communities as "super organisms." By the mid-20th century that expression had been replaced by the more moderate term "ecosystem." Nonetheless, the resonance with lawfulness of individual organisms remained. Odum and Barrett, for example, point out in their classic text *Fundamentals of Ecology* (2005) that in the development of ecosystems there can be found "many parallels in the developmental biology of individual organisms… (as) they progress from 'youth' toward 'maturity.'"

As is easy to imagine, much research went into the study of ecological succession in the 20th century and continues on today. Hundreds of academic articles on this theme are published every year. Why is it so important? This question can lead to a discussion with the students about the urgency of ecological restoration worldwide and how understanding the way ecosystems grow and develop is of great significance in helping heal much of the environmental damage brought about by human beings. (More on this later.)

To help the students grasp the significance of such understanding, introducing them to some form of restoration project in the region where they live is of great value. Hands-on experiences in such projects—combined with background understanding gained in the classroom—can be a huge aid in awakening the students to the process-nature of nature and to the urgent environmental challenges that we face moving into the future.

Regardless of where a school is located, there are certainly, environmental issues related to succession and restoration that can be found nearby. Often the more interesting ones can be embedded in a historical context. In the case of Chicago and other communities located near the Great Lakes, it is fascinating for the students to learn how the scientific study of successions arose along the shores of Lake Michigan in the 1890s.

Although succession has been observed for centuries—for example, when a plowed field is abandoned and allowed to return to forest or prairie—the process had not been studied in detail. That is, until 1896, when Henry Cowles, a graduate student at the University of Chicago, began traveling to the southern shore of Lake Michigan to investigate the Indiana Dunes. These dunes extend some 30 miles along the shore and older, established dunes could be found as far as five miles inland. Dunes of this kind are created when grains of sand blown by the wind begin to gather, forming rapidly changing mounds or ridges. Cowles noticed a striking sequence of changing vegetation as he moved from the youngest dunes that only supported hardy dune grasses and scattered cottonwoods, to older dunes that were populated by plants with deeper, more stabilizing root systems, and by shrubs and pines. Further inward, he found yet older dunes that supported a mixed oak forest with a thick layer of humus. Struck by this sequential development, Cowles began to carefully document the characteristics of each successive plant society, together with changes in soil quality, water availability, wind intensity, and other factors.

What made the dunes such an ideal research project was that Cowles could watch the physical and biological changes in the dune landscape take place almost before his very eyes. In contrast to the formation of a Grand Canyon, or a Yosemite Valley, at the dunes Cowles could drive a stake in the sand and watch how an advancing dune covered it in less than a year! Looking at ecological development as a process unfolding in time, he realized that here, in the rapidly changing dune landscape, he could see spread out in space in the present, a process that had taken place sequentially in climax communities over long periods of time in the past. His observations opened up whole new ways at looking at nature and were a key factor in the establishment of ecology as a science.

At the same time, this perspective raised awareness for the human impact on the environment and—in many instances—its disastrous consequences. As a result, Cowles' way of looking played an instrumental role in the birth of environmentalism in this country. In 1913 Cowles led a group of German scientists on an ecological tour of the United States. When asked what places they most wanted to see on their visit, four locations were most often mentioned: the Grand Canyon, Yosemite Valley, Yellowstone Park, and the Lake Michigan sand dunes. Soon thereafter, Cowles organized a meeting of plant and animal ecologists in Philadelphia that led to the founding of the Ecological Society of America (ESA) in 1915. An offshoot thereof later became the Nature Conservancy. In 1966 Congress authorized the creation of Indian Dunes National Lakeshore. (Chang & Turner 2019; Cowles 1899; Molles 2013; Mullen 1999; Odum & Barrett 2005; Witynski 2021; Worster 1994)

It is very helpful if the teacher is able to provide a historical context for a distinctive ecological setting somewhere in the area where the school is located, and—even better—if it can be visited with students. If this is possible, it creates a much more wakeful interest in concrete ecological investigations in a similar habitat within reach of their school. Since the possibilities for such ecological explorations vary greatly depending on location, only one example of how one might go about this will be given here. This example—which builds on Cowles' study of the Indiana Dunes—can be found in Appendix III.

Microclimates, Microhabitats, Microenvironments

One of the most interesting, close to home, and fundamental aspects of environmental variation can be found in microhabitats. Microhabitats can vary in scale from a few miles to a few yards, and sometimes even reach the scale of a few inches. Such variations in environmental conditions often exist unnoticed in areas quite familiar to the students, so discovering such microhabitats and observing them more closely can be a very eye-opening experience. Since students are usually interested in stories that come out of the teacher's own biography, I often like to place before them the following riddle, one that is quite easily solved. During my teens I lived in Boulder, Colorado. Boulder is located on the eastern edge of the Rocky Mountains. Boulder Canyon runs due west out of the city and cuts deeply into the mountains as it heads for the continental divide. Back in the day, I was always amazed by how different the two sides of the canyon were. As you drive or walk into the canyon, the steep slopes on the left side of the canyon are covered in a dense dark forest of Douglas fir. The right side of the canyon is, by contrast, very savanna like, with ponderosa pines scattered about the steep hillsides, with an understory of dry grasses, shrubs and wildflowers. Two very different worlds can be seen here, face to face and separated only by the highway and Boulder Creek. How is this possible? To the south a dense forest, the modest, high-altitude Western U.S. version of the forests we learned about in the tropics and in the Eastern US, and then, without any transition, a savanna-like landscape.

But, in contrast to what we learned about earlier in the block, we are not witnessing the boundary between two biomes, we are seeing a classic example of two different microclimates in close proximity that are brought about by what is known as "slope aspect." By paying attention to the path of the sun, we realize how it relates very differently to the north-facing and south-facing sides of the canyon. In the course of the year, the north-facing slope (northern aspect) receives much less

Fig. 3.69 Slope effect on coastal sage scrub community, Santa Monica Mountains. (Noah Elhardt. CC BY-SA 2.5 Generic)

direct sunlight, making it significantly cooler and moister (lower evaporation rate) than the south-facing side of the canyon. These contrasting conditions provide the basis for very different constellations of plants, animals and microorganisms living in close proximity to each other.

Once awakened to the influence that slope aspect has on microclimates, it becomes clear that you don't need to go to the Rocky Mountains to discover such habitats—they can be discovered again and again anywhere where you have topographic undulations that create north- and south-facing slopes. Even north- and south-facing sand dunes show variations in microclimate.

If a wooded landscape with steep hills and valleys can be visited with the students, another fascinating aspect of north- and south-facing slopes can be observed: Different growth forms dominate on each slope. Freestanding Linden (*Tilia*) trees, for example, are familiar to the students for the harmonious symmetry of their crowns, which are created by gently curving branches that reach out in all directions.

If we move into the woods, however, and view Linden trees growing on the north-facing slope of a small ravine, we see something very different (Fig. 3.70). The most striking characteristic is the tall, straight, unbranched trunk with a smallish, thin-branched crown located high above. Moving to the south-facing side of the ravine, we get yet another version of a Linden tree. Here we have almost all crown, with multiple crooked, bushy stems instead of a central trunk. Further observation tells us that this is not just an odd characteristic found in Linden trees, but that the same type of variance can be found in oaks, beeches, sycamores, ashes, and so on, when they grow in similar habitats.

Once the students have developed a taste for microclimates/habitats, it can be the beginning of a host of interesting discoveries. If they live in the northern hemisphere, they can observe how the north-facing side of a tree will often grow moss, while the other side will show little or none. Why? Once again: The north side is shadier, cooler and moister than the south side. The mossy side will provide a home for insects, which attract woodpeckers, who dig them out, and so on. Looking around, it soon becomes clear to students that trees and shrubs produce microclimates in the areas that they shade, and that once they die the fallen wood becomes the home of many decomposers: from fungi and bacteria to beetles, ants and termites. The students will often remember how as children they were excited to discover that the underside of stones often harbored a host of organisms that

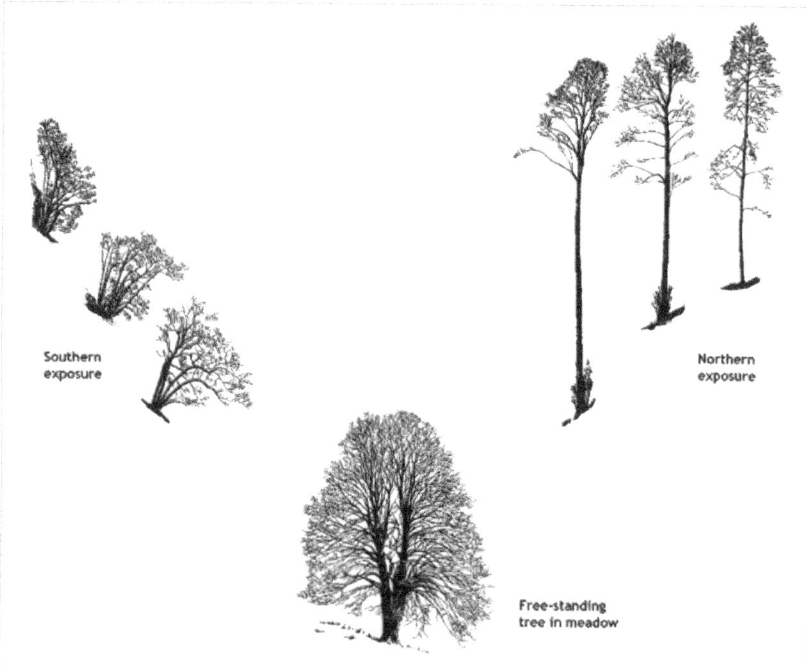

Fig. 3.70 Linden trees on south-facing slope grow crooked and bushy with no main stem. Harmonious symmetry typifies free-standing Lindens. Lindens on north-facing slope show tall, straight, unbranching trunk with small crown. (Drawings from J. Bockemühl [ed.] *Awakening to Landscape*, pp. 36–37, Dornach: Natural Science Section, Goetheanum, 1992)

they rarely saw out in the open. Animal burrows also have their own microclimates, where temperatures are normally more moderate than at the surface. Even dips and mounds that appear across a landscape create significant habitat variances. The dips show higher available moisture and nutrients, higher pH levels, less extreme temperature variations and better developed topsoil (A horizon). On the other hand, they accumulate much more leaf litter, which can hinder seed germination and establishment. And so the story goes, on and on, with much to be discovered and assessed by the students.

Last but not least, it is very illuminating to view the activities of gardeners and farmers, who intentionally create micro-habitats with specific outcomes in mind. An additional question can be: Have these habitats been created with the larger environment in mind, or merely in service of micro-economic interests?

The student's growing awareness of the all-presence of microhabitats and the recognition of the significant influence that human behavior has upon them—conscious or not—can create a natural transition to the direct consideration of the environmental crisis facing humankind in our time and ways in which it has been addressed by various individuals. Biographical insertions here and there of individuals whose idealism and commitment have played an important role in awakening our awareness and understanding of environmental issues can be very inspiring for students at this point in their lives. (Bockemühl 1991,1992; Freeman 2007; Gilliam & Roberts 2003; Holdrege 1996; Holdrege 2013; Lawson, Bennie et al. 2014; Molles 2013; Mutel & Emerick 1992; Suggitt, Gillingham et al. 2011)

Endangered Rainforests: Deforestation, Protection, or Renewal?

The gift of biodiversity and biomass that the rainforests contribute to the biosphere is being seriously threatened. Human-driven deforestation of the Amazon rainforest, for example, has been a major concern for decades. In Brazil, which contains 60 percent the Amazon rainforest, 12 percent (approximately 192 million acres) of that forest—which is an area larger than the state of Texas—has been deforested. From August 1, 2018, to July 31, 2019, for example, 2,400,000 acres were deforested, which amounts to 6,575 acres per day, or 274 acres of rainforest destroyed per hour. This translates into a shocking image that the students will not soon forget: Throughout that year, *2.5 soccer-field-size areas of rainforest (4.5 acres) were destroyed every minute*—never to grow back!

Most of this has been for logging, agriculture and mining purposes. Generally, slash-and-burn techniques have been used. Instead of machetes and axes, the 21st century version of this approach is to use giant tractors and bulldozers with chains to pull down the Amazon's giant trees. After they have dried out for a few months, the trunks are then torched. Hundreds to thousands of acres are taken down and burned in this manner at a time. However, as was noted above, the soil that is left behind is almost devoid of minerals. Although burning releases nutrients from the vegetation, after only a few years of good crop growth on the ash-enriched soil, rains leach the soluble minerals from the soil and everything that remains behind is baked by the tropical sun. Most pasture is degraded and abandoned within several years, which necessitates renewed slash and burn activity in a permanent cycle of rainforest destruction. These sections of the rainforest are lost forever, and with them the amazing diversity of life that they contained.

Another aspect of this rainforest destruction that raises great concern is its impact on the global climate. Rainforests are the earth's largest CO_2 sink. The Amazon alone is estimated to capture up to 25 percent of global CO_2 generation in its plants and other biomass. Overall, it is reckoned that some 250 billion tons

Fig. 3.71 Slash and burn agriculture. (Jami Dwyer. Public domain)

of carbon is stored in tropical forests, which is equivalent to roughly a third of all the carbon stored in the earth's economically recoverable coal, gas and oil reserves. Setting such forests ablaze, the students soon realize, is a double whammy. Not only is an agent lost that absorbs CO_2 from the atmosphere (it is estimated that tropical forests remove 1.3 to 2 billion tons of carbon each year), but the fires release great quantities of CO_2 into the atmosphere again. In the first decade of the 2000s, tropical forest loss emitted around 5 billion metric tons of CO_2 annually, which is more than all emissions from the European Union over the same period. This enormous expulsion of greenhouse gas would have been even more problematic if the remaining uncut tropical forests had not reabsorbed about half of it.

Added to this effect is the loss of water transpiration that comes from rainforest destruction. At first glance, the students do not always appreciate that rainforests not only get rained upon a great deal, but that they also create rain. Amazingly, a single large rainforest tree can move over a thousand liters of water a day from the soil into the atmosphere. Ninety percent of the water that reaches the atmosphere from land is via plant transpiration, and a major proportion of that 90 percent comes from the Amazon basin rainforest—some 20 billion tons of water each day! This compares to the huge quantities of water that flow each day along the Amazon River! Much of the water vapor rising from the trees becomes part of the so-called "atmospheric rivers," which transport water vapor long distances to other areas that need precipitation.

Although the students are well aware of the efforts to reduce carbon emissions stemming from coal, oil and other sources that come from the lithosphere, they often do not realize the pivotal role—referred to above—that the biosphere plays in reducing global warming. In two reports issued by the United Nations Intergovernmental Panel on Climate Change (IPCC) in 2018 and 2019, it was concluded that we need to save the forests in order to stay within an atmospheric carbon budget for only a 2.7°F increase beyond preindustrial warming levels by 2100. They advised that forest loss would need to stop by 2030, and that thereafter wooded areas would need to be increased by up to 85 million acres per year until 2050. A separate analysis concluded that halting tropical deforestation would reduce global emissions by 16 to 19 percent.

Although all forests help, large ones are of the greatest importance for the climate. Despite making up only 20% of the tropical total, intact forests store 40% of the aboveground forest carbon in low latitudes. When they are kept intact, forests deliver a double climate benefit. Not only do they cool the planet by removing CO_2 from the atmosphere, they also cool the local environment through evaporation and transpiration (evapotranspiration).

Forests play a huge role in earth's water household: They retain in their root systems a good part of the daily downpour and absorb the floodwater that streams out of the mountains during the rainy season and when the snow melts. The Amazon catchment area contains almost one-fifth (18%) of all the freshwater that flows into the earth's oceans. If the forests were to disappear, floods and erosion would turn vast regions into desert and make them uninhabitable.

Intact rainforests are, of course, much more than carbon absorbers, environmental cooling agents, and floodwater absorbers. They are the home of complex networks of synergistic relationships between organisms. In the words of Reid and Lovejoy (2022, p. 20):

> *intact rainforests are healthy, complex systems with countless species.... They have predation, pollination, seed spreading, and procreation all happening naturally and in profusion. The have troops, colonies, packs, and pecking orders; micro fauna, megafauna, intrepid migrants and entrenched residents. Harpy eagles eat spider monkeys, tree snakes eat tree frogs, pitcher plants eat ants and ants farm fungus.... Biologists in a large*

Fig. 3.72 Aerial view of the Amazon Rainforest, near Manaus.
(Neil Palmer (CIAT). CC BY SA 2.0)

tropical forest can reasonably hope to find a living thing no scientist has previously encountered. A three-year study concluded in 2018 in Madidi National Park in Bolivia, found 124 new species, including 84 previously unknown plants, 19 fish, 8 amphibians, 5 butterfly species, along with 4 new members of both the mammal and reptile classes.

Biologically, tropical rainforests are the most diverse terrestrial ecosystems in the world. As the earth's largest tropical rainforest, the Amazon rainforest ecosystem is home to about 40,000 plant species, almost 1,300 bird species, 3,000 types of fish, 427 species of mammals, and 2.5 million different insects.

The larger such ecosystems are, the more species they house. In recent years, what are known as "biodiversity hotspots" have been identified around the globe. A large proportion of these endangered areas are found in the rainforests of Africa, Central and South America, and Asia. To qualify as a hotspot, a region must meet two conditions: 1) it must contain at least 1,500 unique higher plant species, and 2) it has to have lost at least 70 percent of the area of its natural habitat. Currently 36 areas qualify as hotspots. Although their intact habitats represent only 2.5 percent of Earth's land surface, they support more than half of the world's plant species as endemics (species found no place else) and almost 43 percent of mammal, bird, reptile and amphibian species as endemics.

> *Big forests are a linchpin in a planetary system... whose intricacies and meaning we don't fully understand. Losing the forest would change more than the reading on the thermometer. Wind, rain, fire, and ocean currents would be rewritten. If we lose too many trees, everything changes. The weave unravels.* (Reid & Lovejoy 2022, p. 28)

(Brienen et al. 2015; Conservation International 2022; Food and Agriculture Organization of the UN 2019; Gaarder 2019; Juniper 2019; Reid & Lovejoy 2022; Rogelj et al. 2019; Rosen 2019; Seymour & Busch 2016; Zaitchik 2019)

From Conservation to Rejuvenation?

Is it possible not only to conserve, but also rejuvenate these forests, these linchpins of the planetary system? What, then, is the difference between conserving and rejuvenating an ecosystem, we ask the students? After some discussion it becomes clear: Conservation usually refers to keeping those influences away that change the rainforest: loggers, miners, and farmers, for the most part. It is about protecting against something—mainly humans. Rejuvenating, on the other hand, includes protection, but goes beyond that. It implies that just keeping destructive influences away is not enough, that aging rainforests could benefit from certain kinds of human activity that would have a renewing influence on them.

Such an idea would have seemed absurd to most 20th century thinkers familiar with the classic literature on Amazonia, which presented it as a lush and bountiful, primal forest that humans merely had to adapt to. Historical ecology of the last 40 years, however, provides us with a very different picture. Using insights from modern ecology, historical ecologists maintain that human activities have been a key factor in shaping biodiversity and environmental health in the Amazon region. What first appeared to be a gift of nature that came about on its own, is now seen as strongly influenced by careful human resource creation and management that has had a significant impact on the diversity and distribution of wildlife there.

In the 1980s surprising discoveries were made that gave rise to a new picture of how Amazonia came about. Clark Erickson from the University of Pennsylvania, for example, found tens of thousands of kilometers of raised banks running across the Bolivian Amazon that had clearly been dug by humans. They had corrugated flooded fields with ridges on which the people living in that area could plant their crops, while also collecting water for irrigation in the dry season. It was the flatland equivalent of the ancient method of terracing hillsides. Today that landscape is essentially empty, but artifacts of human habitation—pottery, charcoal, human bones—remain behind.

Further discoveries followed. A vast system of fish ponds and weirs was found that, together with forest islands lying between them stretched for thousands of square kilometers. Erickson estimates that such fish farming on the edge of the rainforest, combined with the raised fields found there, could have sustained roughly a million people. Buried charcoal uncovered in the roads and mounds suggest that these structures had been build some 2000 years ago.

Such discoveries were, however, only the tip of the iceberg. With further investigation, estimates of the pre-Columbian native population in Amazonia rose to approximately seven million people. The myth of a pristine, largely undisturbed environment prior to the arrival of the Europeans was no longer tenable. The high biodiversity found there was clearly related to past human activities. The wonderfully productive forests—filled with Brazil nuts, palms, liana, bamboo, etc.—had been shaped by indigenous understanding of and interaction with the environment.

One individual, in particular, was pivotal in awakening a new understanding of the profound ways of working with nature that lived in the indigenous peoples of Amazonia for thousands of years. For almost fifteen years, Darrell Posey lived and worked with the Kayapo Indians in Brazil. In the early 1980s Posey mounted an interdisciplinary ethnobiological research project called the Kayapo Project. The project eventually engaged over 30 specialists who possessed very unique knowledge in areas of soils, plants, animals and medicines. Many conferences were then held with scientific and indigenous project members to disseminate the project's findings. Posey formulated one of the central conclusions arising from this work as follows:

> *Native peoples can teach us new models for sustained natural resource use and management. Their ancient indigenous traditions developed through millennia of experiences, observations and experiments are extremely relevant in providing future options for sustainable natural resource management.* (Posey 1992, p. 46)

(Anderson 1990; Erickson 2014; Gomez-Pompa & Kaus 1990; Levis et al. 2018; Morales et al. 2019; Pearce 2005; Posey 1992; Suchantke 1993; Zanotti 2019)

Forest Islands

What does this look like in more detail? The Kayapo are particularly known for their creation of forest islands, or Apêtê, which show the degree to which they can alter and manage ecosystems to increase biological diversity. They begin at small mounds of vegetation created by ant nests in open fields. Small depressions are created that retain moisture and into which seeds or seedlings are planted. After the Apêtês have grown and developed for a time, the Kayapo cut down the tallest trees in the middle, which creates a donut-hole center that allows light into the Apêtê. In this way they create zones that vary in light, shade and humidity. These islands become important sources of medicinal and edible plants. Palms play a prominent role in them, as do shade trees. Full-grown Apêtês look so natural that until recently scientists did not realize that they were human creations. They are common throughout the savannas and wetlands of Amazonia. In Bolivia, archaeologists estimate the presence of some 10,000 such forest islands.

The Kayapo have a refined understanding of how certain plant species develop more vigorously together. They speak of such plants as "good friends." Synergistic plant groups can include dozens of species and necessitate complex patterns of cultivation. Posey compares the way that the Kayapo combine these different "plant energies" to how an artist blends colors when composing a work of art. What appears to a newcomer as a random field of plantings, turns out to have up to five concentric zones, each with preferred varieties of plants that are cultivated in ways best suited to them.

Andreas Suchantke (1993) provides a concrete example of how a Kayapo garden-field with three concentric circles is created and how it is composed. Based on an exact knowledge of the soil makeup and conclusions drawn from the vegetation growing in various locations, a site is chosen for cultivation. The choices are based on the Kayapo's very sophisticated soil taxonomy, which makes horizontal and vertical distinctions based on color, texture, friability, stratification, and drainage qualities. They link such soil qualities to indicator plant species, which allows them to manage those plants accordingly.

The work begins with the removal of undergrowth that is gathered in piles. All trees within a limited area are cut down, with the exception of those that provide important fruits or have medicinal qualities, resulting in a large circular clearing surrounded by trees.

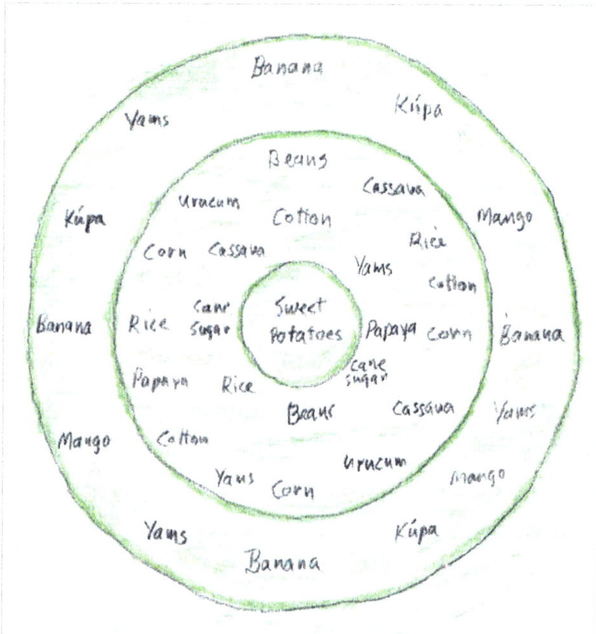

Fig. 3.73 The three concentric areas of Kayapo garden-fields. (Modified after Suchantke)

The *innermost circle* that forms the center of the concentric zones is planted immediately after plowing with sweet potatoes, of which the Kayapo possess 22 varieties. These receive intensive care through continuous fertilizing with wood-ashes, mulching and—amazingly—by continuous burning of small heaps of dry wood that warm the ground just the right amount to further the growth of the sweet potato tubers planted there.

The *middle circle* (ring) is filled with great variety: corn (21 varieties), sweet and bitter manioc (22 varieties), yams, rice, sugarcane, beans, cotton, peanuts (3 varieties), urucu (achiote), papaya, and more. The Kayapo know the needs of each plant type and place them in locations that fit them best—which includes, of course, grouping them with their "good friends." For fertilization purposes, the Kayapo have highly refined composting methods that contain dozens of different ash varieties, which are paired with those plant species best suited to them.

The *outer ring* is populated with long living plants such as different banana cultivars, mango, kupa, and yams. These provide a transition to the surrounding forest. This area is fertilized primarily through mulch and the leaves that fall from the surrounding forest.

The Kayapo also have creative ways to counter the negative effects of certain insects. The leafcutter ants (saúvas), for example, can decimate a planted field in a short time. They are countered by placing nests of the very aggressive Azteca-ants among the plants. When the leafcutters try to enter the Azteca occupied field, they are immediately driven out of with great ferocity.

After four or five years of intensive planting and soil enhancement, the clearing is given up and the slow transformation back to forest begins. But this does not mean that the site is abandoned. Trees are planted in correspondence with the natural succession of the forest. First come quick growing, short-lived trees, which are gradually replaced by taller, long-living species. The seedlings are taken from the forest, with the species selection based on various characteristics: fruit production, medicinal value, usefulness as building material, and so on. A rich herbal undergrowth also develops that attracts various wild animals (agutis, peccaries, pintail deer, pacas). These animals are hunted to provide an important protein supplement to the Kayapo diet, but also to prevent them from grazing the new tree growth too heavily. In this way the rainforest is well cared for. The forest structure is maintained and the makeup only slightly changed—modifications that only an expert would notice.

Another Kayapo practice is the occasional felling of single trees, which creates enclosed, sunlit clearings where hundreds of medicinal herbs seed themselves or are planted. A luxurious ground cover also develops that is absent elsewhere and attracts game.

To sum up: We now know that astonishingly refined methods of agroforestry have existed for a very long time and have been practiced by numerous indigenous peoples in all tropical countries. Not only have they have developed specific methods to meet local conditions, but have continued to refine them over centuries.

This proves concretely that a form of ecologically sound agriculture is possible that does not disturb the fundamental character of the rainforest, and can even renew it. It must be developed, of course, in dialog with nature and hence in harmony with her. But such forms of agriculture are not born of a relationship to nature that is commonly found in the urbanized western world. Ried and Lovejoy (2022) characterize the latter frame of mind well:

> *Saving the megaforests requires more than strategy and tactics. This is a job that needs to be done with feeling. In the modern world, the beings and features of the forests, and even the forests themselves, are commonly*

> *thought of as objects. We humans are separate subjects acting upon them, and nearly all verbs, if we think of this grammatically, are ethically tolerable. Cut, dig, clear, gather, manage, thin, burn.* (pp. 10–11)

This brings us back to Aldo Leopold and his "Land Ethics." Leopold argued that we must learn to see humans and the biological world as part of a single interdependent community, where all members—human and nonhuman—need to be treated ethically. Such a land ethic transforms the role of humans from conquerors to members and citizens of a land-community, where nature is not a means to human's ends, but is honored and valued for its own sake, intrinsically.

For megaforest people such a relationship is nothing new, it is a given. As Kenampa, coordinator of the union of indigenous people in Perus' Javari Valley, describes it:

> *The forest is part of our family. When we look at the forest, we don't just see forest. We see lives. Lives that need us just as we need them.* (In Ried & Lovejoy 2022, p. 133.)

Based on their extensive interactions with occupants of the earth's megaforests, Reid and Lovejoy (pp. 8–9) have come to the following conclusion:

> *To keep Earth livable, we humans must grow out of our habit of transforming woods into landscapes of grass, shrubs, dirt, and pavement. To start, countries can support forest peoples who steward vast areas of the megaforests, Indigenous peoples—whose cultures, spirituality, and practical survival are tied to the trees—control around a third of intact forests. They're the ones who know the woods best.*
>
> *In Canada, Indigenous peoples are reasserting control over their ancestral territories, largely with the support of the government, which recognizes the original peoples as teachers and partners in taking care of nature. Reinforcing these trends is a practical and ethical way to save the megaforests.*

Is this something that could actually still happen, or is it just a fanciful notion? This is an important question for the students. It is vital for them to hear that such thoughts are not just wishful thinking, but are realistic possibilities in the years ahead. (Erickson 2014; Hecht 2007; Morales et al. 2019; Pearce 2005; Posey 1992; Posey 1999; Ried and Lovejoy 2022)

Martin von Hildebrand—The AAA Corridor Project

One outstanding example of how such ideas can bear fruit can be found in the story of Martin von Hildebrand, an anthropologist who has devoted his life to helping indigenous people gain control of their rainforest homelands. Born in New York City, his family moved to Columbia when Martin was five. In 1972, he went as a young man down to the Amazon with a canoe, where he spent several months on his own. Moving along the waterways, he kept meeting other cultures, none of whom could speak Spanish and each of which was living in traditional ways that were completely different from the Western world he had grown up in. He was fascinated and deeply touched by what he met there. Then, suddenly another side of Amazon-reality burst in upon him: He began running into rubber camps where the indigenous people were basically enslaved and being forced to work from 3:00 in the morning until 10:00 at night. They had been there as forced labor for years, as had their parents before them. Von Hildebrand was shocked. He soon learned that the children of these forest dwellers were taken by force at the age of six to missionary schools and told that their culture—ancestors, beliefs and daily life—was wrong. He felt devastated that the Western culture he had grown up in could treat these people in such a way.

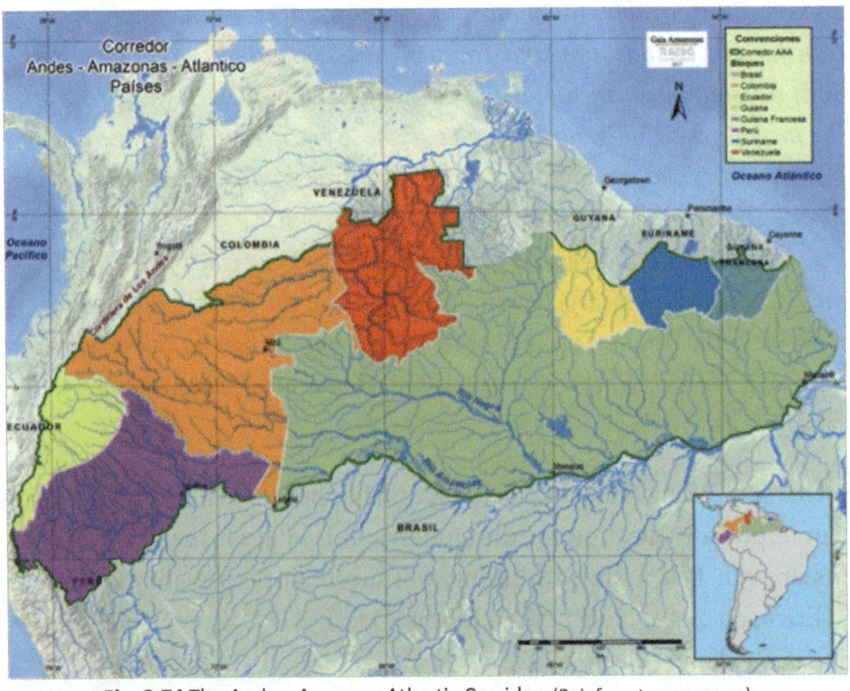

Fig. 3.74 The Andes, Amazon, Atlantic Corridor. (Rainforestconcern.org)

This was the turning point in his biography. He decided to devote his life to fighting for the rights of indigenous peoples, to stop such attacks on their way of life, and to help them regain their land and their culture, and free their children. Since that awakening, some 50 years ago, von Hildebrand has helped indigenous communities take back control of millions of acres of Amazon rainforest. To start with, he spent the rest of the 1970s living in communities of forest Indians, learning about their culture, ways of thinking, and gaining their trust. During the 1980s he worked for the government and became close to two successive Columbian presidents, two whom he told stories about the Indians and their communities. With their help, he managed to get indigenous land rights—and their broader rights in general—recognized. In 1990 he created an organization called Gaia Amazonas that helped these communities make the most of their new rights. Today they run their own schools in their own languages and have their rights and governments recognized in the national constitution. They now control an area of the Columbian rainforest larger than Great Britain and are able to negotiate with national and regional governments as equal partners.

What does this mean for the rainforest. In von Hildebrand's view:

> *The indigenous people have developed a world view that aims at safeguarding the forest, not only from a spiritual point of view but because their lives depend on it. It is their home, their identity, their understanding of themselves and the world. They are one with the rainforest. They are the rainforest. In Colombia they offer better protection for the rainforests than the National Parks.* (Juniper 2019, p. 201)

But von Hildebrand's vision does not stop there. Looking beyond Columbia at the larger Amazon basin, he began an initiative known as "the Andean, Amazon, Atlantic Corridor Project" (AAA Corridor Project, Fig. 3.74). In this he partnered with indigenous communities, NGOs, and national governments in the northern Amazon region. Stretching from Columbia across Venezuela and Brazil to the Atlantic Ocean—and thereby creating the world's largest ecological corridor—the project aims to safeguard 333 million acres (521,240 sq. miles) of forests. Nearly 80% of the corridor already exists as a patchwork of protected areas and indigenous territories. Beyond this mosaic of areas, Columbia, Brazil and Venezuela have been working to create protected areas adjacent to the corridor that could be included in it. This would bring the corridor to an area of approximately 495 million acres (773,000 sq. miles). It could even be expanded beyond that to include other countries such as Guyana, Surinam, French Guyana, and Peru, which have also expressed interest in participating in the project. This would expand the corridor to approximately 655 million acres (1,023,000 sq. miles). (Gaia Amazonas 2018; Juniper 2019; Rainforest Concern 2015; Reubold 2015)

But, as von Hildebrand emphasizes, it is important to keep in mind:

> *It's not just a protected area. We're not creating a park. It's a laboratory. We are creating a laboratory for developing responses to climate change with both indigenous and Western knowledge.* (in Reubold 2015, p. 2)

The Future of the Rainforests

What do other individuals who are very familiar with and have been working on these issues for many years have to say about them? Take Tony Juniper, for example, the chair of England's governmental conservation agency. Juniper is someone who has been engaged for over 30 years in helping conserve the earth's tropical forests and is able to formulate a very down-to-earth picture of where things are heading. How does he see the future?

At the conclusion of his 400+ page book: *Rainforest – Dispatches from Earth's Most Vital Frontlines*, Juniper gives us the following picture of what's ahead:

> *The fact that I can paint this relatively optimistic picture of the road ahead says a great deal about how far the world has come, and in a very short time. When I began my campaign work for the rainforests back in 1990, I wouldn't have thought such progress would be possible. In the early 2000s, and the time of that demonstration in Davos, political and business will was either absent or weak. Today, however, the case for the rainforests has been successfully made and action is being taken, albeit not yet on a sufficiently large scale, or with the consistency of political leadership that is needed.*
>
> *Having suggested that a corner is being turned, it has to be said the new road ahead is a long one. But going down it as quickly as we can will be vital for protecting not only the future of the tropical rainforests for their own sake, but also that of people. Should we drive hard and fast, we could secure the bargain of the century—a task yielding massive returns from what is in the greater scheme of things a modest investment….*

> *We know all we need to know to do this. What it requires now is for all of us who are in a position to help to join in with this historic task. This is not only a call to action for company executives and government ministers, but for all the rest of us, in looking carefully at who we vote for, what we eat and consume.... After all, there's no reason why we must continue to watch the inexorable decline of the tropical rainforests. We can save most of what is left and put back much of what's gone, if we want to.* (Juniper pp. 397–398)

Such positive developments are important for students to hear about as their ecology block comes to close. They are at an age when their awakening—or already well-developed—idealism needs to see that there are goals, big and worthwhile goals, that their generation—and they as individuals—can support and help come to fruition.

Resources Part III

Ahrens, C. (2012). *Meteorology Today*. Pacific Grove, CA: Brooks Cole.

Alexandra, J. (1996). *Mephistopheles' Anvil*. Spring Valley, NY: Rose Harmony.

Anderson, A. (1990). "Deforestation in Amazonia." In Anderson, ed. *Alternatives to Deforestation*. New York: Columbia University Press.

Anderson, R. (2006). "Evolution and origin of the central grassland of North America." *Journal of the Torrey Botanical Society* 133(4): 626–647.

Axelrod, D. (1985). "Rise of the grassland biome, central North America." *Botanical Review* 51 (2): 163–201.

Barlow, J. (2003). *Weather extremes shed light on prairies past*. News Bureau, University of Illinois. ps://www.uiuc.edu

Bockemühl, J., Schad, W., and Suchantke, A. (1978). *Mensch und Landschaft Afrikas*. Stuttgart: Verlag Freies Geistesleben.

Bockemuehl, J. and Jaervinen, K. (2006). *Extraordinary Plant Qualities for Biodynamics*. Edinburgh: Floris Books.

Bockemühl, J. (1985). "Pflanzengestalt und Lichtverhältnisse." In Endlich, ed. *Der Organismus der Erde*. Stuttgart: Verlag Freies Geistesleben.

________. (1991). *Die Landschaft als Organismus*. Die Drei (7/8): 547-580.

________. (1992). *Awakening to Landscape*. Dornach: Natural Science Section.

Boucher, D., et al. (2014). *Deforestation Success Stories: Tropical Nations Where Forest Protection and Reforestation Policies Have Worked*. Union of Concerned Scientists.

Bray J. (1960). "The composition of savanna vegetation in Wisconsin. *Ecology* 41: 721–732.

Breckle, S. (2002). *Walter's Vegetation of the Earth*. Heidelberg: Springer.

Brienen, R., et al. (2015). "Long term decline of Amazon carbon sink." *Nature* 519: 344–348.

Briggs, J., Knapp, A., Blair, J., Heisler, J., Hoch, G., Lett, M., and McCarron, J. (2005). "An ecosystem in transition: causes and consequences of the conversion of mesic grassland to shrubland." *Bioscience* 55 (3): 243–254.

Bronstein, J. (2015). *Mutualism*. Oxford: Oxford University Press.

Chadwick, O. (1995). "What good is a prairie?" *Audubon* Nov.-Dec., 35–46, 114–116.

Chang, C. and Turner, B. (2019). "Ecological succession in a changing world." *Journal of Ecology*. 13 February.

Clements, F. (1916). *Plant Succession*. Washington.

Collins, S., Knapp, A., Briggs, J., Blai,r J., and Steinaur, E. (1998). "Modulation of diversity by grazing and mowing in native tallgrass prairie." *Science* 280: 745–747.

Conservation International (2022). *Biodiversity Hotspots*. Conservation.org

Cornell Compost. http://compost.css.cornell.edu/

Cowles, H. (1899). "The ecological relationships of vegetation on the sand dunes of Lake Michigan," *Botanical Gazette* 27: 95–117. Also reprinted in Real L. and Brown J., eds. (1991). *Foundations of Ecology, Classic Papers with Commentaries*. Chicago: University of Chicago Press.

Cribb, J. (2006). "Fly Times When the Dry Comes." *The Australian*.

Cronon, W. (1983). *Changes in the Land*. New York: Hill & Wang.

Curtis, J. (1959). *The Vegetation of Wisconsin*. Madison: University of Wisconsin Press.

Dey, D. (2021). "Restoration of Midwestern Oak Woodlands and Savannas" (PDF).

Diaz, S. and Cabido, M. (1997). *Plant Functional Types and Ecosystem Function*.

EcoLogic Development Fund (2018). *Slash and Burn Agriculture*. www.ecologic.org

Elliot, J. (1973). *Botany*. London: The English Universities Press.

Erickson, C. (2014). "Amazonia: The historical ecology of a domesticated landscape." In Hecht et al., eds. *The Social Life of Forests*. Chicago: University of Chicago Press.

Food and Agriculture Organization of the United Nations. (2019). "Flying rivers – how forests affect water availability downwind and not just downstream." July 15, 2019.

Freeman, J. (2007). *Science 101: Ecology*. Irvington, NY: HarperCollins.

Gaarder, B. (2019). "The highest Amazon deforestation rates in over a decade." Regnsdogfondet.

Gaia Amazonas. (2018). "AAA Corridor: A macro-regional inititative to face climate change. www.gaiaamazonas.org

Gichoki, H., Gakahn, C., and Mwange, E. (1996) "Savanna ecosystems." In McClanaham and Young, eds. *African Ecosystems and Their Conservation*. Oxford: Oxford University. Press.

Gilliam, F. and Roberts, M., eds. (2003). *The Herbaceous Layer*. New York: Oxford University. Press.

Gimmingham, C. (1951). "The use of life form and growth form in the analysis of community structure as illustrated by a comparison of two dune communities." *Journal of Ecology* 39(2): 396–406.

Gitay, H. and Noble, R. (1997) "What are functional types and how should we seek them?" In Smith, T., Shugart, H., and Woodward, F. *Plant Functional Types*. Cambridge: Cambridge University Press, 3–19.

Golley, F. (1993). *A History of the Ecosystem Concept in Ecology*. New Haven: Yale University. Press.

Gomez-Pompa, A. and Kaus, A. (1990) "Traditional management of tropical forests in Mexico." In Anderson, ed. *Alternatives to Deforestation*. New York: Columbia University. Press.

Gore, A. (1992). *Earth in Balance—Ecology and the Human Spirit*. Boston: Houghton Mifflin.

Graham, L., Graham, J., and Wilcox (2006). *Plant Biology*. Upper Saddle River, NJ: Pearson.

Greenwalt, C. (1990). *Hummingbirds*. New York: Dover.

Grohmann, G. (1989). *The Plant*. Biodynamic Association.

Grzimek, B. (1979). *Grzimeks Tierleben – Insekten (Band 2)*. Munich: DTV.

Hargrove, E. (2003). "Weak anthropocentric intrinsic value." In Light A. and Rolston H., eds. *Environmental Ethics—an Anthology*. Malden, MA: Blackwell, 175–190.

Harry, D. (2018). "Red-headed Woodpecker." Madison Audubon. https://madisonaudubon.org

Hecht, S. (2007). “Kayapos Savanna Management: Fire, soils and forest islands in a threatened biome.” In W. Woods, ed. *American Soils.* Berling: Springer Verlag.

Heinrich, B. and Bartholemew, G. (1979). “The ecology of the African dung beetle.” *Scientific American* 241 (5).

Heinrich, B. (2012). *Life Everlasting.* New York: Houghton Mifflin Harcourt.

Heisler, J., Briggs, J., Knapp, A., Blair, J., and Seery, A. (2004). “Direct and indirect effects of fire on shrub density and aboveground productivity in a mesic grassland.” *Ecology* 85 (8): 2245–2257.

Holdrege, C. (1996). *Genetics and the Manipulation of Life.* Hudson, NY: Lindisfarne Press.

_______ (2013). *Thinking Like a Plant.* Great Barrington, MA: Lindisfarne Books.

Holdrege, M. (2010). *From Beach to Savanna.* Ann Arbor: ProQuest.

Husemann, F. and Wolff, O. (1982). *The Anthroposophical Approach to Medicine.* Spring Valley, NY: Anthroposophic Press.

Imes, R. (1990). *The Practical Botanist.* New York: Simon & Schuster.

Johnson, H. (2022). *Rainforest.* National Geographic Society.

Johnson, M. (2002). *American Woodland Indians.* Oxford, GB: Osprey Publishing.

Julius, F. (1969). *Metamorphose.* Stuttgart: JC Mellinger.

Juniper, T. (2019). *Rainforest.* Washington, DC: Island Press.

Kassa, Y. and Tekalign, W. (2022). “The population size and distribution of diurnal large wild mammals in the southern Great Rift Valley, Ethiopia. *Scientific World Journal*, Feb. 27.

Katz, E. (2003). “Is there a place for animals in the moral consideration of nature?” In Light A, Rolston H. eds. *Environmental Ethics—An Anthology.* Malden, MA: Blackwell, 85–94.

Kirt, R. (2000). *Prairie Plants of the Midwest.* Champaign: Stipes.

Knapp, A., Blair, J., Briggs, J., Collins, S., Hartnett, D., Johnson, L., and Towne, E. (1999). “The keystone role of bison in North American tallgrass prairies.” *Bioscience* 49 (1): 39–50.

Koenig, W. and Knops, W. (2002). "The behavior ecology of masting in oaks." In McShea and Healy, eds. *Oak Forest Ecosystems: Ecology and Management for Wildlife*. Baltimore: Johns Hopkins University. Press, 129–148.

Koepf, H. (1990). *Bio-Dynamic Agriculture. An Introduction*. Hudson, NY: SteinerBooks.

Kump, L., Kasting, J., and Crane, R. (2009). *The Earth System*. Upper Saddle River, NJ: Pearson Prentice Hall.

Lawson, Bennie, et al. (2014). "Topographic microclimates drive microhabitat associations at range margin of a butterfly." *Ecogeography*. Vol. 37 (8), 732–737.

Leach, Mark K. and Givnish, Thomas J. (1988). "Identifying highly restorable savanna remnants." In *Transactions of the Wisconsin Academy of Science, Arts and Letters* 86: 119–128.

Leopold, A. (1949). *A Sand County Almanac*. New York: Oxford University Press.

Leuthold, C. (1998). "Oekogenese – die Biographie eines Landschaftsorganismus." In Suchantke, ed. *Oekologie*. Stuttgart: Verlag Freies Geistesleben.

Levis, C., Flores, B., et al. (2018). "How people domesticated Amazonian forests." In *Frontiers in Ecology and Evolution* 5: 71.

Light, A. and Rolston, H. (2003). "Introduction: Ethics and environmental ethics." In Light, A. and Rolston, H., eds. *Environmental ethics—an anthology*. Malden, MA: Blackwell, 1–12.

Madson, J. (2004). *Where the Sky Began—Land of the Tallgrass Prairie*. Iowa City: University of Iowa Press.

Maglianesi, M., Blüthgen, N., Böhning-Gaese, K., and Schleuning, M. (2014). "Morphological traits determine specialization and resource use in plant–hummingbird networks in the Neotropics." *Ecology* 95(12).

Makweti. (2020). "Beyond the Ball." www. makweti.com

Manning, R. (1995). *Grassland*. New York: Penguin.

McHugh, T. (1979). *The Time of the Buffalo*. Lincoln: University of Nebraska Press.

McNaughton, S. (1985). "Ecology of a grazing ecosystem: The Serengeti." *Ecological Monographs*. Volume 55, Issue 3, 259–294.

McPherson, G. (1997). *Ecology and Management of North American Savannas.* Tucson: University of Arizona Press.

McShea, W. and Healy, W. (2002). "Oaks and acorns as a foundation for ecosystem management." In McShea and Healy, eds. *Oak Forest Ecosystems: Ecology and Management for Wildlife.* Baltimore: Johns Hopkins University Press, 1–9.

Minnesota Department of Natural Resources (2006). "Tomorrow's Habitat for the Wild and Rare: An Action Plan for Minnesota Wildlife, Comprehensive Wildlife Conservation Strategy." Division of Ecological Services.

Molles, M. (2013). *Ecology—Concepts and Applications.* New York: McGraw-Hill.

Moore, A. (2019). *Amazon Rainforest Fires.* Raleigh: NC State College of Natural Resources.

Morales, J., et al. (2019). "Historical landscape domestication in ancestral forests with nutrient-poor soils in northwestern Amazonia." *Forest Ecology and Management* 446.

Mullen, W. (1999). "Birth of a Movement." *Chicago Tribune.* April 2009.

Mutel, C. and Emerick, J. (1992). *From Grassland to Glacier—the Natural History of Colorado.* Johnson Books.

Nardi, J. (2007). *Life in the Soil.* Chicago: University of Chicago Press.

Nash, R. 1989. *The Rights of Nature—A History of Environmental Ethics.* Madison: University of Wisconsin Press, 290 pp.

National Park Service (2016). "Bison Bellows." *Bison Stories—Bison* (U.S. National Park Service), nps.gov

National Wild Turkey Federation (2022). "All for the Oak: Makeover of the Michindoh." nwft.org

Nelson, P. (2017). "Oak savannas." *MNopedia.* Minnesota Historical Society.

Nuzzo, V. (1986). "Extent and status of midwest oak savanna: presettlement and 1985." *Natural Areas Journal* 6: 6–36.

New York City (2009). NYC Compost Project Tip Sheet. www.nyc.gov/wasteless/compost

Oak Savannas (2021). "Oak savannas: characteristics, restoration and long-term management." Oak Savannas.org

Odum, E. and Barrett, G. (2005). *Fundamentals of Ecology*. Belmont CA: Thomson Brooks/Cole.

Onishi, N. (2022). "To fight climate change, Canada turns to indigenous people to save its forests." *The New York Times*, Nov. 16.

Palmer, C. (2003). "An overview of environmental ethics." In Light, A. and Rolston, H., eds. *Environmental Ethics—An Anthology*. Malden, MA: Blackwell, 15–37.

Pearce, F. (2005). *Deep Jungle*. London: Transworld Publishers.

Pigliucci, M. (2001). *Phenotypic Plasticity*. Baltimore: Johns Hopkins University Press.

Pommeresche, H. (2020). *Humusphere: Humus, a Substance or a Living System?* Greeley, CO: Acres USA.

Posey, D. (1990). "Forest islands, Kayapo example." *Overstory* #34.

_______ (1999). "Traditional knowledge, conservation and 'The Rainforest Harvest.'" In Plotkin and Famolare, eds. *Sustainable Harvest and Marketing of Rainforest Products*. U.S. Department of Energy.

Pritchett, W. and Fisher, R. (1987). *Properties and Management of Forest Soils*. New York: John Wiley & Sons, 494 pp.

Rainforest Concern (2015). "Andes-Amazon-Atlantic Corridor Project." rainforestconcern.org

Raven, P., Evert, R., and Eichhorn, S. (2004). *Biology of Plants*. New York: W.H. Freeman.

Raven, et al. (2012). *Biology of Plants*. New York: McGraw Hill.

_______ (2016). *Biology*. New York: McGraw Hill.

Reichman, O. (1987). *Konza prairie*. Lawrence, KS: University of Kansas Press.

Reid, J. and Lovejoy, T. (2022). *Ever Green*. New York: W.W. Norton.

Reubold, T. (2015). "Martin von Hildebrand: An audacious plan for the Amazon." Earth Journal Network.

Rogelj, et al. (2019). "Mitigation pathways compatible with 1.5°C in the context of sustainable development. Climate Change and Land. *An IPCC Special Report*.

Rolston, H. (2003). "Value in nature and the nature of value." In Light, A. and Rolston, H., eds. *Environmental ethics—an anthology*. Malden, MA: Blackwell, 143–152.

Rosen, J. (2019). "The Amazon rainforest is on fire." *Los Angeles Times*. Aug. 26.

Sands, R. and Summer, W. (2017). *Growing Sustainable Children – A Garden Teacher's Guide*. Great Barrington, MA: SteinerBooks.

Savage, C. (2004). *Prairie*. Vancouver: Greystone, 308 pp.

Savanna Oak Foundation, Inc. "Ecology of savanna plants." https://www.oaksavannas.org/savanna-ecology.html

Schad, W. (1985). "Vom Geist der Natur—Lebenskreis und Lebensumkreis des Mondhornkaefers." In Endlich, B., ed. *Der Organismus der Erde*. Stuttgart: Verlag Freies Geistesleben.

_______ (2020). *Threefoldness in Humans and Mammals*. Hillsdale, NY: Adonis Press.

Schaller, G. (1975). *The Serengeti Lion—A Study of Predator-Prey Relations*. Chicago: University of Chicago Press.

_______. (1983, 2013). *Golden Shadows, Flying Hooves*. New York: Alfred A. Knopf, Inc.

_______. (2007). *A Naturalist and Other Beasts*. San Francisco: Sierra Book Club.

Schaumann, W. (1982). "Die Bildung der Pflanzenqualitaet." In Schad, ed. *Goetheanistiche Naturwissenschaft*: Botanik. Stuttgart: Verlag Freies Geistesleben.

Schroeter, C. (1989). "Alpenblumen." Cited in Grohmann, *The Plant*, Vol I. 93.

Seymour, F. and Busch, J. (2016). *Why Forests? Why Now? The Science, Economics and Politics of Tropical Forests and Climate Change*. Washington DC: Center for Global Development.

Shiva, V. (2016). *Who Really Feeds the World? Failures of Agribusiness and the Promises of Agroecology*. Berkeley: North Atlantic Books.

Sobrig, O., Medina, E., and Silva, J. (1996). "Determinants of tropical savannas." In Sobrig, O., Medina, E., and Silva, J., eds. *Biodiversity and Savanna Ecosystem Processes—A Global Perspective*. Berlin: Springer, 31–44.

Sokolow, J. (2003). *The Great Encounter.* Armonk, NY: M.E. Sharpe.

Stevens, W. (1995). *Miracle under the Oaks: The Revival of Nature in America.* New York: Pocket Books.

Stokes, D. and Stokes, L. (1989). *The Hummingbird Book.* Boston: Little Brown & Co.

Struzick, E. (2022). "Canada's First Nations Protect Millions of Acres of Their Lands." *Yale Environment* 360.

Suchantke, A. (1982). *Der Kontinent der Kolibris.* Stuttgart: Verlag Freies Geistesleben.

_______ (1993). *Partnerschaft mit der Nature.* Stuttgart: Urachhaus.

_______ (2001). *Eco-geography.* Great Barrington, MA: Lindesfarne.

Suggitt, Gillingham, et al. (2011). "Habitat microclimates drive fine-scale variation in extreme temperatures." *Oikos.* Vol. 120 (1), 1–8.

Talbott, S. (2010). "Context matters – the epigenetics revolution. In *Context* #23.

Thomson, B. (1977). *The Changing Face of New England.* Boston: Houghton Mifflin.

_______ (1977). *The Shaping of the American Heartland.* Boston: Houghton Mifflin.

UCMP (2021). *The Tundra Biome.* Berkley.edu

Uhl, et al. (1990). "Studies of ecosystem responses to natural and anthropogenic disturbances." In Anderson, ed. *Alternatives to Deforestation.* New York: Columbia University Press.

Van Jaarsveld, E. (2000). "Welwitschia mirabilis." *Veld & Flora* 86: 176–179.

Volk, T. (2003). *Gaia's Body—Toward a Physiology of Earth.* Cambridge: MIT Press.

Von, Heyer G. (1976). *Der Regenwurm.* Hamburg: Eigenverlag.

Warren, K. (1990). "The power and promise of ecological feminism." *Environmental Ethics* 12/2: 125–146.

Warren, K. and Cheney, J. (2003). "Ecological feminism and ecosystem ecology." In Light, A. and Rolston, H., eds. *Environmental ethics—an anthology.* Malden, MA: Blackwell, 294–305.

Weitzel, E. (2021). "The ecology of the first Thanksgiving." *Scientific American.* Nov. 22, 2021.

Whitney, G. (1996). *From Coastal Wilderness to Fruited Plain.* Cambridge: Cambridge University Press.

Wildlife Online (2021). "Deer Food & Feeding – Gut Flora & Fauna." www.wildlifeonline.

Winckler, S. (2004). *Prairie.* Iowa City: University of Iowa Press, 132 pp.

Witynski, M. (2021). *What Is Ecological Succession.* Uchicago.edu

Wolf, J., Kriss, K., et al. (2021). "Gut microbiome composition predicts summer core range size in two divergent ungulates." *FEMS Microbiol Ecol.* 2021 April 13;97(5): fiab048.

Worster, D. (1994). *Nature's Economy.* Cambridge: Cambridge University Press.

Zaitchik, A. (2019). *Rainforest on Fire.* The Intercept.

Zimmerman, L. and Molyneaux, B. (1996). *Native North America.* London: Macmillan.

Appendix I.
Human Embryology

In what follows the reader will find examples of material that I have found effective when teaching a block on human embryology. At certain junctures along the way, I attempt to step back and characterize the dynamic involved in the phenomena under consideration. Such characterizations—which in the classroom would take place with the help of the students—are intended to stimulate in the students a qualitative and will-permeated experience going well beyond a mere spectator-like "knowing of the facts." To achieve this, it is extremely helpful to engage the students in artistic activity such as drawing, modeling, and bodily movement. Success depends on turning the phenomena into "happenings" that speak to and call forth the students' inner activity.

Gametogenesis

An egg cell (oocyte) is spherical, a shape contrasting strongly with the specialized form of most cells. Moreover, egg cells develop in isolation—each enclosed in an ovarian follicle—whereas most cells are integrated into tissues and organs. The solitary egg cell, so large that it can be seen with the naked eye, is essentially a ball of cytoplasm. Indeed, it contains so much cytoplasm that it cannot survive for long in this condition once it leaves the incubating conditions of the ovary. Those that eventually enter the uterine tube do so in a very open and vulnerable condition. A six-month-old female fetus contains around six million egg primordia (primordial oocytes), of which only about two million are still alive at birth. These diminish continually in number, so that at puberty only around 200,000 are left. This number goes on shrinking until none is left at menopause.

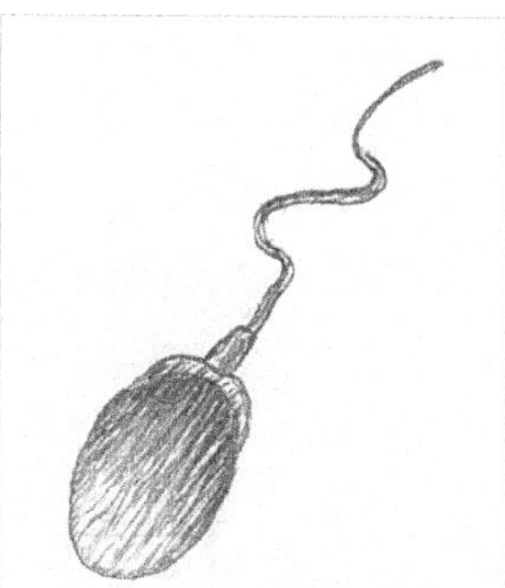

Fig. I-1 Sperm cell.

Whereas an egg cell develops far more cytoplasm relative to its nucleus than any other cell in the body, sperm cells do just the opposite. They reduce the amount of their cytoplasm to such a degree that their nuclei are essentially naked, covered only by a very thin protein layer. The extreme difference between these two cell-types can be seen in the fact that a sperm cell contains only 1/100,000 of the cytoplasm found in an egg cell.

Sperm are more linear in shape, extremely small, and anything but vulnerable. They are able to swim through acids and can be frozen to -60°C (-75 degrees Fahrenheit) for years; after thawing out, they become active again as though nothing had happened. Whereas eggs are moved passively through the medium they are in, sperm swim actively through their environment. In contrast to egg cells, which are produced before birth and find themselves in a continuous dying process, the sperm cells appear only at puberty, and from then on produce at a phenomenal rate of about 80 million per day (approaching 1,000 per second!) in a healthy male. A single ejaculation produces 150–250 million. This enormous, never-ending productivity stands in great contrast to the long-ago-produced egg primordia, and their ever-declining numbers.

Fertilization

Having differentiated into such polar opposite, fragmentary half-cells, each with only a portion of what makes up a complete cell, these gametes need each other. Their wholeness can be restored only through fertilization. Of the millions of spermatozoa deposited in the female genital tract, only 300 to 500, swimming upstream through the uterus and oviduct, will actually reach the egg cell.

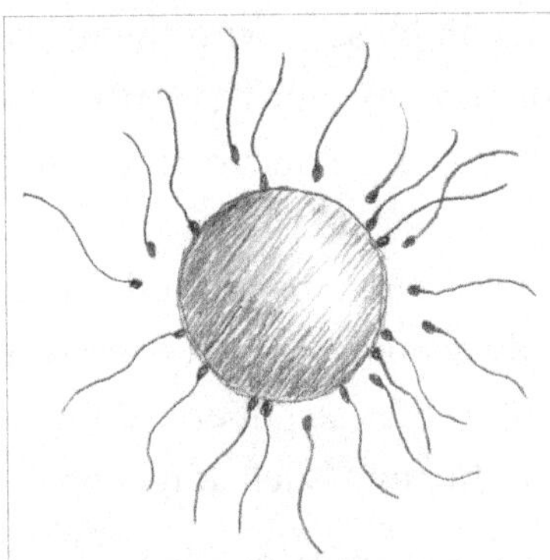

Fig. I-2 Artistic rendering of sperm cells arriving at egg.

The sperm surround the egg with their heads against the membrane of the oocyte and their tails pointed outward. The tails then begin to move rhythmically in a synchronized manner and the egg begins to rotate. For a few hours, this dialogue, this dance before fusion, known as the "preconception attraction complex," takes place. Depending on this interaction between the two cells, conception will or will not take place. If conception is successful, the cell membrane of one sperm cell merges with the cell membrane of the egg and the contents of sperm and egg fuse with one another. Existing until now with an extreme cytoplasmic emphasis (egg), on the one hand, and a one-sided nuclear emphasis (sperm), on the other, these polarities have now united. They have come together because they cannot continue to exist without each other. Having survived on the brink of death only through the assistance of other cells, this union of the gametes allows a whole series of processes to unfold that lead to a significant increase in the metabolic/respiratory activity of the ovum. This, in turn, makes it possible for the ovum to discard its surrounding entourage of hitherto life sustaining support cells.

What arises through this conjunction, however, is more than the mere reestablishment of a cellular whole. The fertilized ovum (zygote) is not just an ordinary cell; it is a whole new organism! Having passed through an almost lifeless stage, the genetic material is now completely rearranged, chromosomes repaired, and the cytoplasm with its organelles has gone through a renewal process.

Characterization: After presenting new material such as described above, the task then is to characterize it so that it touches the students' feeling life and engages their will-based imagination. In the case of gametogenesis and fertilization, the phenomena themselves speak so powerfully that the teacher need only guide the discussion in order for the students to experience vividly the amazing contrast between the two cell types and the significance of their meeting.

On the one hand, we have cells that are generated before birth, exist isolated in egg follicles, and continually die away without replacement until all are gone and infertility results. On the other hand, we have a volcano of productivity (close to 1,000 sperm per second) that surges into life with teaming masses reminiscent of an enormous school of microscopic fish. Eggs are huge, almost choking on the enormous amount of cytoplasm they contain. They are so massive and immobile that they must be transported by the medium in which they find themselves. Sperm are lean and mean, containing only 1/100,000 of the cytoplasm found in eggs. They are like overtrained, undernourished athletes, who swim tirelessly upstream, totally consuming themselves in the process. As a result, of the millions of sperm that may enter the female genital tract, only a few hundred survive the journey all the way to the egg.

Both sperm and egg are helplessly one-sided and doomed to die. Despite their desperate conditions, when the two meet, it is anything but a drastic, intemperate event. Indeed, for hours during the rhythmical dance of the sperm and ovum, it is unclear whether they will actually unite. What has seemed so one-sided in isolation is now very measured and balanced as they prepare to merge. Conception is not just a mixture or balancing of extremes, but through it a whole new level of existence comes into being (what Goethe would call *Steigerung*). This is a new creation, one with the potential to bring forth an entire organism!

If successful, discussions of this nature will help the students come to a vivid experience of the concept of polarity and the higher synthesis that can come from it. This is the seed of what Rudolf Steiner calls a living concept, one that can live on in them, one that can grow and develop with them. In the context of this block, it will have awakened in them a sensitivity to such processes and will help them develop "eyes" with which to enter into and experience more vividly the dynamic that unfolds as we pursue the further stages of human embryonic development.

The First Week

After the fusion of egg and sperm, the haploid nuclei unite to form chromosome pairs. As it drifts slowly toward the uterus, the unicellular zygote divides into two cells after about one day. These cells continue to divide again and again within the next few days, at which point—because of its similarity to a mulberry—the organism is called a morula. By the end of the first week, it has reached approximately 120 cells, but with no growth yet in size!

As it enters the uterus some three days after conception, the cluster of blastomeres undergoes a process of compaction, which segregates to some degree inner cells from peripheral cells while forming a more dense structure. At four days, fluid begins to penetrate through the zona pellucida, forming a central cavity (blastocoel). The embryo is now known as a blastula (or blastocyst) and begins to develop in two polar directions. Toward the center, the so-called embryoblast develops, while the outer layer, the trophoblast, unfolds peripherally.

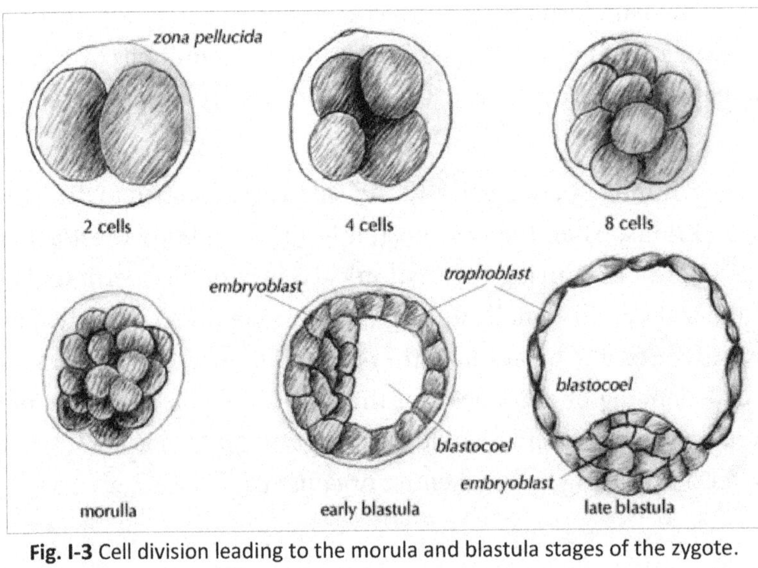

Fig. I-3 Cell division leading to the morula and blastula stages of the zygote.

Characterization: During the first week after conception, we see a serene, straightforward start to life: The egg drifts along, rhythmically dividing. It can be helpful at this point to inquire of the students whether human structures—those composed of smaller units, like bricks—come about in a similar manner? The students easily recognize that egg development is not an additive process, in which parts are simply put together. What we began with, the fertilized egg, is a whole, a whole organism that has begun to divide itself into sub-units known as cells. In fact, during the morula phase, each of these sub-units has the potential to develop into a whole organism.

During this first week of development, we have seen the zygote drift toward the uterus, isolated from direct contact with its surroundings through the impermeable zona pellucida, while continually dividing within the contained boundaries it sets. Through this process, the one-sided plasma-dominance in the egg cell is gradually balanced out. This series of cell divisions (cleavages) leads to smaller and smaller subdivisions (blastomeres) of the fertilized egg. In this way the original plasma amount is allocated among an ever-growing number of nuclei, thereby overcoming the one-sidedness we found in sperm and egg and leading gradually to a normal nucleus-to-plasma ratio. At a certain point, cell division without growth must cease, or the cytoplasm-to-nucleus ratio would once again become extremely one-sided. The chance to grow arises when trophoblast enzymes soften the zona pellucida to a degree that it breaks open (hatching) and the embryo is able to absorb nutrients, grow, and open itself to its environment once again.

As it grows more and more compact, we see the beginnings of a first differentiation (polarization) in the blastula, with the trophoblast cells unfolding in the periphery (expansion) and the embryoblast cells concentrating toward the center (contraction).

The Second Week

At the transition from week one to week two, the process of implantation begins. Freed from its isolation within the zona pellucida, the embryo begins—with the help of enzymes that it produces—to nestle itself into the mother's uterine wall. At the same time, the mother organism must "make space" for this nesting (nidation) process, allowing a stranger to grow in her own body. Depending on the nature of this dialogue between the mother as host organism and the embryo, implantation will succeed or fail. If successful, the embryo will find itself embedded within the maternal tissue, with the uterine wall completely closed, by the tenth day. During implantation, the trophoblast expands at an enormous rate, going from approximately 100 cells at seven days, to about 1,000 at eight days, and reaching around 10,000 cells by day nine! Space for this enormous unfolding process must be provided by the mother organism.

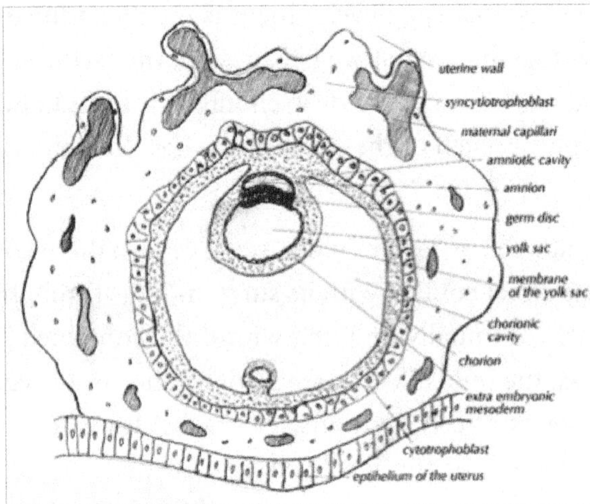

Fig. I-4 The embryo on day 13.

The centrally-located embryoblast, by contrast, expands very slowly, developing only two fluid-filled vesicles: the amnion cavity and the yolk sac. The future body of the embryo—the embryonic disc (germ disc)—will develop where the surfaces of these two vesicles meet.

Viewed from above, the flat embryonic disc is round, thus exhibiting radial symmetry. It contains two layers, and is often referred to as the bilaminar, or germ disc. It stands in strong contrast to the rapidly expanding and much larger trophoblast that surrounds it on all sides. The trophoblast plays a pivotal role during this phase of the embryo's development. Also consisting of two layers (the synctiotrophoblast and the cytotrophoblast), it performs all the functions (nutritional, respiratory, elimination, defense, etc.) that the organs of the embryo's body-to-be will perform in the future.

Characterization: Comparing the gestures of the first and second weeks, we find in the transition from one to the next a huge shift. Whereas in week one we may have felt a gesture of contraction with smaller and smaller subdivisions without growth and within an isolating boundary (the zona pellucida), in week two we find a gesture of expansion in the opening up to, and growing together with, the mother organism through the trophoblast. This emergence from isolation is coupled with a tremendous outward expansion and growth (from 100 to 10,000 cells in two days!). The expanding periphery merges with the mother's tissue (the endometrium layer of the uterus) and takes on all the life functions of the organism. Seemingly left behind in the center is the tiny two-layered bilaminar disc. One huge transformation in one short week! In addition to emphasizing the signature polarity created by the trophoblast and embryoblast during the second week, it can also be noted that this has been a "week of twos." Not only has the trophoblast separated into two layers (the cyto- and synctioblast) during this week, but the embryoblast has differentiated into the epiblast and hypoblast, and around those two layers contrasting cavities have formed: the amnion and yolk sac.

The Third Week

During the second week the trophoblast expanded dramatically into the periphery, whereas the embryoblast in the center grew little. As development continues, the focus of activity changes. At the beginning of the third week the circular embryonic disk begins to grow longitudinally toward the future head (cranial) pole, thus creating a head-base (cranial-caudal) dimension and the beginnings of a right-left (bilateral) dimension.

At the caudal (base of the spine) end of the embryonic disc, a small indentation soon appears (the primitive streak) that grows toward the head pole. Through the indentation cells force their way under the upper, epiblast layer and then sideways between it and the lower, hypoblast layer.

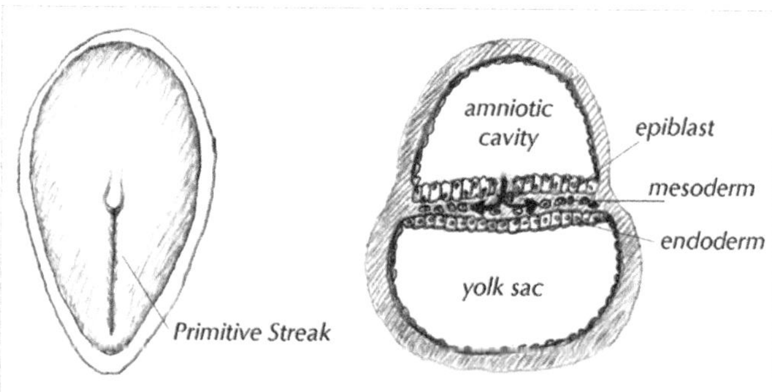

Fig. I-5 Germ disc. Gastrulation process.

This gastrulation process establishes fully the right-left dimension and bilateral symmetry of the embryonic disc and leads to the formation of the notochord. The notochord lies in the exact center of the embryo and establishes the central axis, as well as the dorso-ventral (back-front) dimension. Through this invagination process a third layer has been created between the hypoblast and epiblast layers. We can

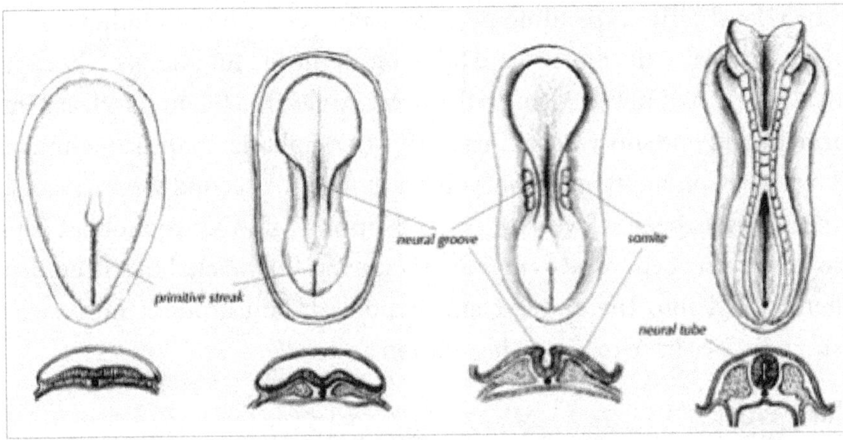

Fig. I-6 Formation of the neural tube.

now speak of three germ layers: the ectoderm (from the epiblast), endoderm (from the hypoblast) and mesoderm, the new third layer. Soon a rhythmic subdivision creates the somites that lie next to the notochord. These will later disintegrate into embryonic connective tissue (mesenchyme). The elongated embryonic disc is now growing rapidly, particularly at the head pole. In the middle of the cranial-caudal axis a second, neural groove begins to form, which then elongates toward both ends of the embryonic disc. This neurulation process results in the formation of the neural tube—the primordium of the entire nervous stem.

At the "heart" of this transformation is a development that has yet to be described: the formation of the circulatory system. Early in the third week, the vascular system begins to develop in the form of cell clusters (blood islands) in the mesoderm along the outside of the yolk sac membrane. Cavities form in these clusters. Gradually, as the clusters grow they connect and their cavities merge, forming capillaries. As they become longer, the capillaries extend all the way to the embryonic disc. Some of the cells in the capillary walls separate off and form blood cells, which begin to flow within the capillary fluid: A circulation of blood connecting center and periphery has begun! Coming from the periphery, this blood flows toward the head, where it is dammed up by denser tissue and must flow back again toward the caudal end of the disc and then out into the periphery. The vessels that turn back soon merge into networks and form the heart primordium — which is located in front of the future head. Around the 20th day, the newly formed heart begins to pulse! With this, the embryo has developed its first organ system, an organ system that mediates between its own developmental processes at the center and its placental functions in the periphery.

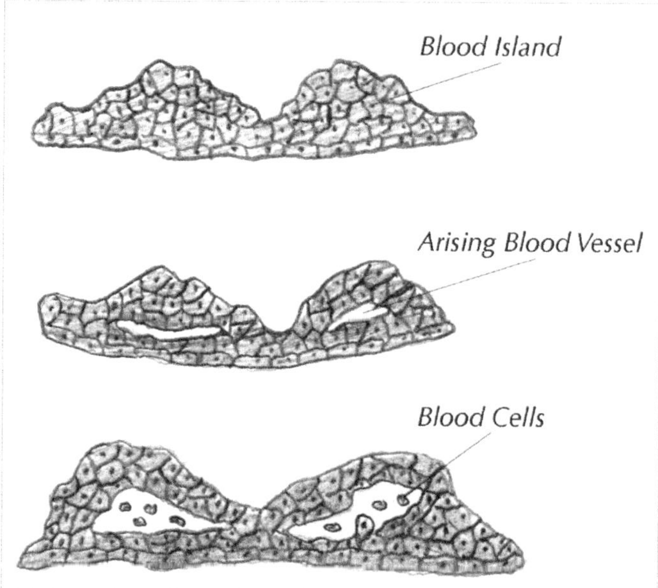

Fig. I-7 Formation of blood vessels and blood cells.

Looking back on the third week, we see a process of internalization taking place. What had been peripheral and undifferentiated as organ functions is now taken inside and begins to incarnate, first in the germ layers and then in organ primordia. With the development of the circulatory system, center and periphery are linked in a new way, and with the heart a true center is found, a center that becomes the basis for consciousness.

Characterization: Yet again we find a huge shift in developmental emphasis. From the periphery the focus now moves to the center, but in a wholly new way. Until now, radial and spherical symmetry had been the organizing principles in all-new structures. In the third week we see a new spatial orientation arising: The embryonic disc begins to lengthen, creating a head-base polarity and bilateral symmetry. Even more dramatic are the invagination processes (gastrulation and neurulation) that create for the very first time an interior layer within the blastocyst.

Periphery and embryoblast now begin to interact in a new way, as well. The blood vessels and blood cells that have formed flow into the embryoblast, only to be dammed up by the head pole and sent back again, but this time permeated by the rhythmical activity of a new organ—the heart. Rhythm from within begins to permeate the organism for the first time! When we reflect upon the central role that rhythms of the most varied kind play in the human organism, we realize what a milestone this is. What we have seen as expansion and contraction as a developmental dynamic from the very outset (for example in sperm-egg, embryoblast-trophoblast) now takes on a temporal dimension in the systole-diastole of the heart.

What we have been witnessing is the outset of a multilevel invagination process of great significance. It marks the beginning of an internalization process, which will gradually draw functions, formerly housed in the periphery (trophoblast) of the organism, into a three-dimensional body. In weeks to come, these will further differentiate into a system of interacting internal organs.

Discussing these developments with students often leads to the realization that internalization is not only to be found on a morphological level, but also in regard to our own inner capacities. Often people and events we meet or learn about in the course of our lives do not remain external to us; they become part of our biography and individual identity when they are internalized in the form of memories, values, capacities.

The Embryonic Period: Fourth to Eighth Week

At the beginning of the fourth week a dramatic new development begins: The germ disc starts folding in on itself in two different dimensions: lengthwise (cephalo-caudal) and sideways (laterally). This involves a dramatic expansion of the amnion cavity, which allows it to enclose the embryonic disc on all sides. As part of this in-folding process, the heart moves from its location in front of the head into the chest region. At the caudal end of the disc, the enveloping gesture of the amnion moves the connective stalk toward the ventral center, which crowds it together with the yolk sac stalk and gives rise to the primitive umbilical chord.

During this process the head grows more rapidly than the base of the spine, as is evident when one compares the two ends of the embryo. At the head pole we find the first faint beginnings of the eyes and ears. At the caudal pole, by contrast, leg buds are beginning to appear.

With the massive curvature of the dorsal surface of the embryo, the ventral side becomes enveloped and now lies within. With the head at one end and the developing legs at the other, the embryo has found its center in the newly formed heart and umbilical cord.

Early in the fifth week, the embryo rolls itself even more tightly together, so that head and base of the spine are almost touching. The back is very rounded with the head leaning over the large heart protrusion. The limbs are paddle-like.

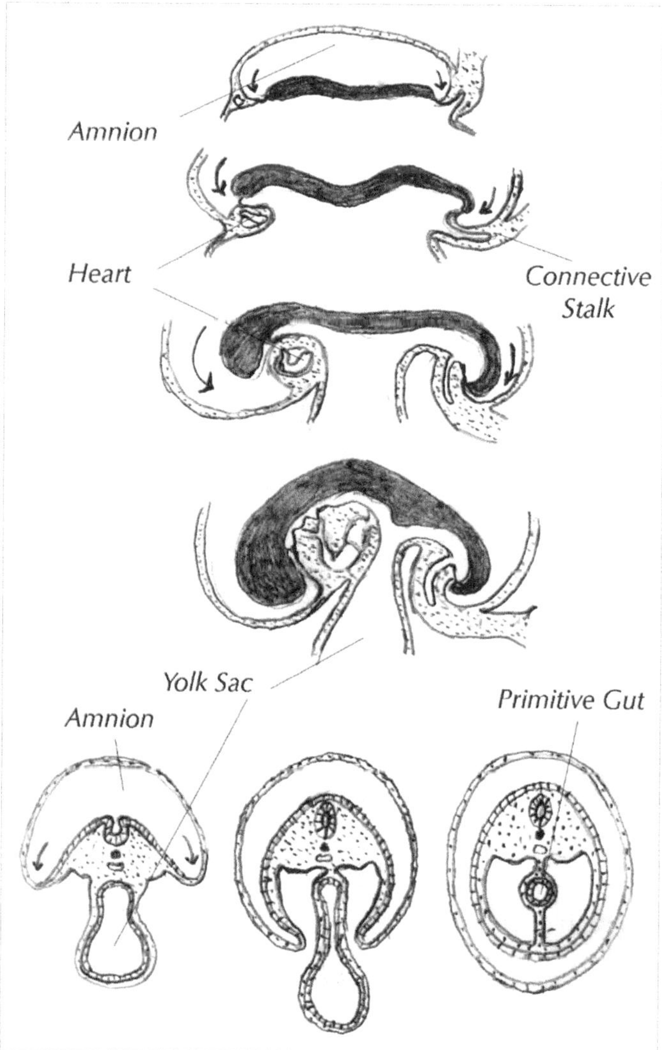

Fig. I-8 The folding of the embryonic disc.

With the sixth week we have a major shift in gesture: The enfolded embryo begins to straighten and unfold. The spine is less rounded and beginning to stretch. The heart has grown larger, and the digestive tract—with too little space in the abdomen—creates a bulge in the umbilical cord. The limbs are now longer and the hand digits are becoming visible.

In the seventh week the head has become rounder, ears and eyes are noticeable, the heart has been absorbed into the torso. The limbs have grown longer, fingers and toes are more evident.

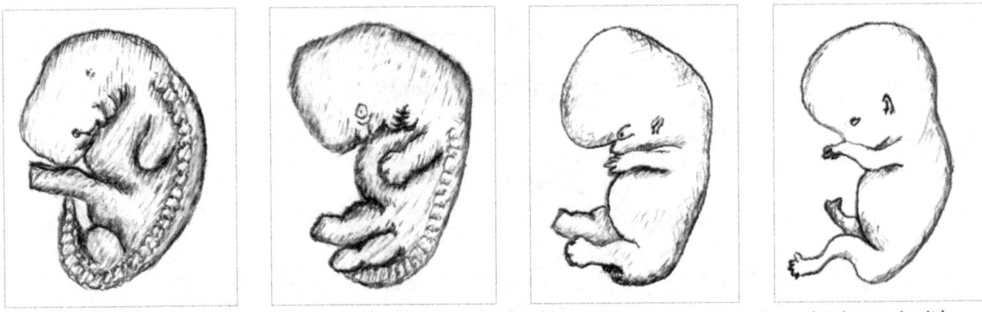

Fig. I-9 Unfolding of empryo form 5th and 6th weeks (R) and further in the 7th and 8th weeks (L).

In the eighth week, the 3cm. long embryo has developed a large forehead; the neck, waist, and hips are forming, the legs are now located under the torso. During this week fingers and hand, toes and feet become fully formed. Not only that, but the arms rotate outward and the legs inward, which allows the palms to turn upward and the feet to turn downward.

Beginning already in the fourth week and continuing through the eighth week—that period when the embryo rolls up to an extreme (head and base of the spine facing each other) and then gradually unfolds again—ectoderm, mesoderm, and endoderm give rise to the organs and structures of the human body. What had been peripheral functions—in the trophoblast without specific organs—has been internalized during this phase of development through the formation of large organ systems: brain, spinal cord, nerves, cartilaginous skeleton, musculature, heart with atria and ventricles, liver with two lobes.

During this process, the nervous system has ascended cranially (morphological cerebralization) and the viscera have descended to a more caudal position. In between (in the upper thorax) the heart and lungs have been given space to unfold. During the same period, the head has drawn back (influenced by the rapid growth of the brain) and the limbs have lengthened and grown forward (ventrally).

Characterization: Students are at first surprised by this dramatic in-folding of the embryo, but on further reflection it begins to make more sense. Is it not, after all, a gesture we have seen before—but now at a much higher, more differentiated level? Indeed, this developmental dynamic—which creates a spherical, isolating form that shows its back to the world (antipathy)—may remind us, for example, of the egg during its first week: enclosed in the zona pellucida and focused entirely on internal cell division. The alternating (and simultaneous) dynamic of expansion and contraction has been a formative component of the entire developmental process, from gametogenesis until now. Thus it should not surprise us that one last contraction—in which head and base of the spine roll together to enclose the developing heart—should prepare for a final, gradual unfolding that will lead to uprightness and the uniquely human relationship to the world around.

Expanding our reflections further afield, students are often reminded of phenomena such as a caterpillar withdrawing into its cocoon, thereby creating an isolated interior space allowing for the development of entirely new organs, and with these the ability to engage the world in an entirely new way. Further musing reveals that this gesture is familiar in yet other ways. For example: the thinker who seeks to withdraw into an inner space of quiet thought when focus is needed for serious problem solving.

Fig. I-10 Rodin: *The Thinker*. Metropolitan Museum of Art. (CC0 1.0 Universal public domain dedication)

Many teenagers will even recognize themselves in a variation of this gesture, when, for example, they shut the door to their rooms and proclaim: "Keep out, I want to be alone!" in order to come to terms with tumultuous feelings and emerging questions about life and the world.

Indeed, just as the caterpillar eventually brings forth the butterfly, the thinker new thoughts, and the teenager centeredness and inner direction, so too does the embryo open up once more, beginning in the fifth week. In the embryo, this unfolding manifests in the gesture of uprightness — the uniquely human posture that will lead eventually to our bi-pedality and vertical linearity, in which the head rests on the trunk and spine, and the arms are free for creative activity.

Fetal Period

The fetal period is characterized by rapid body growth and maturation of the aforementioned organ systems. Striking is how head growth slows down relative to the rest of the body. In the third month it is roughly half the entire body length, in the fifth month it measures about one-third, and at birth the head is about one-fourth of the entire body length. It is also noteworthy how, during the last two months of gestation, the previously soft, almost transparent body of the embryo gradually gains density and weight through mineralization.

Many additional deliberations can follow at this point in the block, depending on the class and the time available. For example, the development of various organ systems, such as the complex transformations of the embryonic heart and brain, can be considered in more detail. Phenomena that lead to an understanding of the uniqueness and developmental openness of the human organization, such as heterochrony and juvenility can be productively explored through comparisons with the developmental phases of various mammals.

"Recreating in the wake of ever-creating nature"

All in all, the success of this block hinges on the students' ability to engage their own will-permeated imaginative faculties as they learn about human embryological development. To do justice to that development, it is necessary to recreate it—to the best of one's ability—in one's own imagination. In doing so, we practice a modest version of what Goethe described as "recreating in the wake of ever-creating nature." To the degree that the students can recreate these embryological gestures in their own picturing activity, to that degree these developments will lose their abstract—merely conceptual—character and begin touch the students at a deeper level.

Tools to aid in this process are plentiful in a Waldorf school. Careful, esthetically rendered drawings call forth, of themselves, a "recreative" process that engages the feelings and helps create a will-based connection with the topic at hand. Movement exercises in class can raise into awareness many of the qualitative gestures at work in developmental processes. If an art teacher can be engaged for sculptural exercises in conjunction with this block, additional deepening opportunities will be available.

Resources

Appenzellar, Kaspar (1976). *Die Genesis im Lichte der menschlichen Embryonalentwicklung*. Basel: Zbinden.

Barfield, Owen (1967). *Romanticism Comes of Age*. Middletown, CT: Wesleyan University Press.

Bie, Guus van der (2001). *Embryology*. Driebergen: Louis Bolk Institute.

Blechschmidt, Erich (2004). *The Ontogenetic Basis of Human Anatomy*. Berkeley: North Atlantic Books.

Gelder, Tom van (2012). *Phenomenology*. www.dynamisch.nu.

Goethe, Johann Wolfgang von (1988). *Goethe: Scientific Studies*. Miller, D., ed. New York: Surkamp.

Rohen, Johannes (2007). *Functional Morphology*. Hillsdale, NY: Adonis Press.

Sadler, T.W. (2000). *Langman's Medical Embryology*. Baltimore: Lippincott, Williams & Wilkins.

Schad, Wolfgang (1982). *Die Vorgeburtlichkeit des Menschen*. Stuttgart: Mellinger.

Selg, Peter (2000). *Vom Logos Menschlicher Physis, Bd. II*. Dornach: Verlag am Goetheanum.

Steiner, Rudolf (1924). *Young Doctors Course* (Lecture 1). GA 316. April 21, 1924.

________. (1921). *Man's Being, His Destiny, and World Evolution* (Lecture 2). GA 226. May 7, 1921.

________. (1919). *The Study of Man* (Lecture 9). GA 293. August 30, 1919.

Suchantke, Andreas (1975). *Zur Menschenkunde des Reifealters*. Erziehungskunst, Jan. 1975, 4–15.

Van der Wal, Jaap (2010). *Embryo in Motion*. DVD. www. portlandbranch.org.

Verhulst, Jos (2003). *Developmental Dynamics*. Ghent, New York: Adonis Press.

Wilmar, Frits (1975). *Vorgeburtliche Menschenwerdung*. Stuttgart: Mellinger.

Appendix II.
Colorado High Country—a Comparison

In Southwest Colorado's San Juan Range, between Lake City and Silverton, the mountain sides are dominated by two tree species very different in nature and appearance: Engelmann spruce (*Picea engelmannii*) and quaking aspen (*Populus tremuloides*). Since there are roughly 60,000 tree species on earth (approximately 600 of which are conifers), it may surprise at first that only two of these species appear preeminent in that area. Ranging from small patches to swaths that cover hundreds of acres, they carpet the mountain valleys and ridges from an altitude of 9,000 feet right up to the tree line at approximately 11,000 feet. Each species builds very large stands. The stands often exist side by side, although transition zones are not infrequent.

The Comparative Method

One way to approach this striking phenomenon that is fitting to the topic and also has educational value, follows a central idea developed by Rudolf Steiner in his lecture course, *The Study of Man* (CW 293). In that work Steiner emphasizes the value of characterizing rather than merely describing and defining. Characterization goes beyond mere description; it first receives pulse and life when one phenomenon is compared with another. Taking an example out of the grade school curriculum, Steiner describes how "the teacher does not merely describe consecutively… but rather places cuttlefish, mouse and man side-by-side and relates one with another. The interrelationships will prove so manifold that the result will not be a definition but a characterization" (*Study of Man*, lecture 9). One key aspect of this method is that the things being compared characterize each other: Phenomenon characterizes phenomenon. The points of view that arise using this approach thus stem to a large extent from the objects themselves—from their similarities and differences—rather than from predispositions of the observer.

It is certainly no accident that Rudolf Steiner constructed the foundations of Anthroposophy and his own approach to natural science in direct connection with a master practitioner of the *comparative method*: Goethe. In the so-called *Light Course* (CW 320, lecture 1), Rudolf Steiner describes how "Goethe considered all of that which is known as the natural scientific method to be merely a tool that enables one to group sensory phenomena in such a way that they themselves reveal (speak about) their secrets." Goethe characterized his approach to nature as *delicate empiricism*. By that he meant a careful attentiveness to the phenomena, a kind of receptive "listening" that silences the premature urge to form an explanatory hypothesis and allows the phenomena to gradually "speak for themselves" (Naydler, 1996). With these thoughts as a background, this essay explores the Spruce-Aspen treescape of high-country Colorado using such a comparative approach. If successful, it will allow the two tree species under consideration to speak for themselves and characterize each other—as well as the environment in which they live.

General Impressions

When you enter a stand of Engelmann spruce high in the Colorado Rockies, you are struck immediately by the quiet and serenity that envelops you. A mood of almost stoic silence is tangible in the shade of these giant trees; the air is cooler than elsewhere; the ground is soft beneath its carpet of needles. Moving a few feet over into a stand of quaking aspen brings a very different experience. You come out of the cool darkness into a world of light and movement. A mood of sanguine levity pervades the area. You not only feel the breeze as it moves through the trees, but see it, too, as the leaves tremble and twirl with its passage.

To advance from such general impressions to a more wakeful awareness of what distinguishes these two neighboring stands requires looking at various characteristics of each tree in the same sequence, thus allowing—through their juxtaposition—each tree to speak not only of itself, but of the other as well. In this way a more holistic picture of each species and its environment can gradually arise.

Fig. II-1 Engelmann Spruce [*Picea engelmannii*].
(Jeremiah LaRocco. CC BY-SA 2.0 Generic)

Shape and Color, Leaves and Light

Beginning with the overall shape and color of the two species, one is struck by the spire-like verticality of the Engelmann spruce (Fig. II-1). The trunk forms a central column, compared to which the sideward extending limbs seem secondary and subordinate. The side branches are astonishing thin compared to those of most trees. They extend horizontally in whorls, like spokes of a wheel. Each year they produce a new whorl nearly identical to the previous one. The shortest and uppermost branches extend out at a slightly upward angle, whereas the longer, lower branches begin to angle downward through their own weight. The linearity that informs the overall architecture of the spruce extends right into the leaf formation. The leaf-blades have been reduced to four-sided, 5/8ths to one-inch long, tip pointed (acicular) needles that are arranged spirally along the twig where they are anchored (Fig. II-2).

Fig. II-2 Englemann Spruce twig with many narrow leaves.
(Steven J. Baskauf. CC BY-SA 4.0 International)

Each 2–3 inch twig can carry 50–100 new-growth needles each year. These are added to those that have remained on the tree from the past 7–10 growing seasons The total number of leaves that such a tree carries is phenomenal! Due to their prevalence and the close-knit way they lie along the stem, these blue-green leaves absorb a great deal of light, thus creating the shadowy darkness that one experiences within and around these tightly woven trees. In this habitat, very little undergrowth is found.

Whereas the Engelmann spruce is narrow at the top and widens as it nears the earth, quaking aspen, with its limbless lower trunk, is narrow near the earth and widens first in the upper regions of the tree. Although it also unfolds around an upwardly-striving central trunk, its ascending limbs extend further than those of Engelmann spruce and create an open, airy, and rounded crown that is anything but dark (Fig. II-3). Because the lower branches die and fall away (prune) as soon as the limbs above them begin reducing the available sunlight, the radiant white trunk is completely open and visible in the lower regions of the tree. Not only does the bright white bark add luster to its surroundings by reflecting back much of the sunlight, but it also transforms some of the light as well.

Fig. II-3 The open crown of a Quaking Aspen [*Populus tremuloides*]. (Photo by Joseph O'Brien, USDA Forest Service, Bugwood.org)

Whereas bark in most trees represents a lifeless outer periphery that isolates the tree from its surroundings, the bark of quaking aspen contains a chloroplast-containing green layer inside the outer white one. With this living boundary, the quaking aspen not only enters into a substance-transforming life-process with the sunlight, but through the vitality of this layer it also opens itself up as a food source for elk and numerous insects. The open crown of the quaking aspen does not create dark shadows like the dense network of needle-like spruce leaves, but appears to "play" with the light through the almost incessant trembling and pulsing of its roundish, finely toothed leaves, which are carried out into the surrounding air on slender, flattened stalks. (Compare figures II-4 and II-5 with figures II-1 and II-2).

Fig. II-4 Quaking Aspen leaves.
(Šarūnas Šimkus GNU Free Documentation License.)

Fig. II-5 Light-permeated Quaking Aspen grove.
(Zion National Park. CC BY-SA 2.0 Generic)

Fig. II-6 Aspens during autumn. (Famartin. CC BY-SA 3.0 Unported.)

This light-airy quality reaches its zenith in the autumn, when the leaves turn spectacular tints of gold and yellow before they wither and fall as part of their yearly, sun-determined rhythm (Fig. II-6).

Not only the aspen themselves thrive on direct sunlight, but their open, dynamic crowns allow much light to pass through to the benefit of other life forms below. In most aspen groves a thriving understory of shrubs, forbs, and grasses is to be found, with wetter sites showing an abundance of wildflowers. They also provide an important wildlife habitat. Young sprouts are heavily browsed by deer and elk, while the rapidly decaying trunks of soft aspen wood provide refuge for many cavity-dependent species.

Habitat, Growth, Age and Size

Engelmann spruce are shade-loving trees that do not readily establish in the open. They prefer, instead, 40–60% of full shade and moist seedbeds. Their early growth is surprisingly slow at higher elevations. After five years the average height of an Engelmann spruce is just 1–3 inches(!), and at age ten it is not unusual to find seedlings only 6–8 inches tall. During the next ten-year period they grow more rapidly, reaching an average height of 4–5 feet by the time they reach age 20. Under the heavy shade of a closed canopy, however, it is not uncommon to find 80–120 year old trees that have only reached a height of 3–5 feet. Engelmann spruce is, nonetheless, a long-living tree. It matures at about 300 years and often lives to be 450—although 500-600-year-old trees are not unknown. On favorable sites they often develop into large, 100–150 feet tall trees, with trunk diameters that can reach as much as six feet.

Quaking aspen, by contrast, thrive in sunlight, where they grow very rapidly. When conditions are right, young trees can grow five feet annually for the first several years. Most of their growth takes place during the first 20 years, after which it slows down considerably. Quaking aspen usually top out at about 75 feet when conditions are good, and usually only live to around the age of 80.

Root System

Whereas the shallow root system of Engelmann spruce grows about five inches in the first year, and gradually expands with the growth of the tree thereafter, the lateral roots of quaking aspen reach six feet or more already in the second year. Curling and tangling beneath the surface, these cord-like lateral roots meander and undulate for great distances without tapering. Along the way they form buds that may lie dormant for years. If however, open ground and sufficient sunlight are available, a new aspen shoot (ramet) will suddenly grow up from the root system below. Although sustained initially by the parent tree, a ramet develops up to 15 feet of new main-roots within ten weeks of sprouting. This manner of vegetative reproduction from a single individual leads to whole communities of aspen that are genetically identical and thus referred to as clones.

Flower and Fruit

The flowering regions of Engelmann spruce and quaking aspen can also be compared. Although they both produce staminate and pistillate flowers, Engelmann spruce is monoecious, producing both on the same tree (seed cones in the upper crown, pollen cones in the lower), while quaking aspen is dioecious, with the two types of flowers appearing on separate plants. The seed cones of Engelmann spruce are originally erect, but become weighty and pendent (downward hanging) after pollination. In fact, the pistillate cones of these trees lignify to such a degree that one can say, with Grohmann (1974), that they are made of wood! Each cone appears like a small tree, with its scales grouped in spirals around the central spindle (Fig. II-7).

Fig. II-7 Lignified Pistillate Cone of Engelmann Spruce. (Susan McDougal. USDA Plants.)

Fig. II-8 Loosely hanging catkins of Quaking Aspen. (Famartin. CC BY-SA 4.0)

The substances and processes that manifest outwardly in the formation of colorful blossoms in most plants remain "withdrawn" in the Engelmann spruce, appearing only in the aromatic fragrances of resin and essential oils in the needles. The seeds dispersed from these cones are equipped with a single wing that extends from one end of the seed. These winged seeds rotate as they fall through the air, with the denser seed body located earthward below the wing. This enables them to pass between small objects and penetrate more deeply into the forest floor. Although these seeds are sometimes carried by the wind, they do not generally travel far from the source.

The pistillate flowers of quaking aspen develop quite differently. In early spring the loosely hanging (pendent) catkins look like pinkish caterpillars (Fig. II-8). After pollination, they develop into tiny capsules that soon release great numbers of extremely light seeds (150,000–250,000 per ounce) that are buoyed by tufts of long silk that allow the wind to carry them for miles before they pile up like cotton balls on the forest floor.

In contrast to spruce cones—that fall intact to the ground after releasing their seeds—the catkins of quaking aspen appear to dissolve into the mountain air around them. Whereas seed production can begin already at age two or three in quaking aspen, Engelmann spruce only become seed bearing between the ages of

15 and 40. Quaking aspen can already produce large seed crops between the ages of 10 and 20, while Engelmann spruce produce their most abundant crops first when they have reached an age of 150–250.

The salient features discussed above can be more easily overviewed when placed side-by-side in tabular form.

In Summary

On the one hand, we find a tightly-knit, dark spire that needs (and also creates) shade and moisture, while producing countless stem-embracing, light-absorbing, needle-like leaves. It brings forth cones that grow heavy, pendant, and turn to wood. It produces progeny that, although they grow extremely slowly, live for centuries and reach great size. On the other hand, we have an open-crowned, light-loving tree, whose shimmering leaves extend out in active interplay with the surrounding light and air. Its flowers dissolve in countless minute seeds that spread for miles in the wind. These trees spring up in light-filled openings, grow rapidly in their early years, but are displaced and disappear at a relatively young age. They do not lose their potential for renewal, however, for beneath the surface their ubiquitous root system waits for open, sun-permeated conditions (created by fire or some other form of disturbance), that will allow them to shoot up exuberantly and build, once again, their youthful, diversity-engendering community. Whereas the taciturn and somber communities of Engelmann spruce live on and on with grandfatherly perseverance, quaking aspen presents us with short-lived, youthful vitality, coupled with mercurial renew-ability.

The Part in the Whole

Following this descriptive comparison, we can seek for an overriding gesture in each of these trees and in their relationship to one another. It is striking, for example, that for Engelmann spruce everything unfolds slowly—and preferably in the shade—into dense, contracted forms. Spruce forests are dark, taciturn, monotone ecosystems that live on and on if not disturbed by outer forces. With their tightly contracted, linear leaves, they reveal minimal openness to their surroundings, even to the point of "ignoring" the seasons as their leaves live on through light-poor, snow-filled winters. By comparison, quaking aspen unfold rapidly when sufficient light is available, their forms are dynamic, light, and relatively short-lived. Reflecting the change of seasons dramatically with their golden-orange autumn leaves, their openness to their surroundings is expressed in a dynamic-airy canopy and

Engelmann Spruce	Quaking Aspen
Creates a cool, quiet, shady habitat lacking in herbal undergrowth due to darkness and carpet of needles.	Creates an open, light-penetrated habitat with diverse undergrowth.
Dense, narrow, conical crown (narrow at top, wide at base) with the trunk as a central column and short, thin branches that spread in close repetitive whorls.	Single, radiant-white main trunk. The overall form is expansive above, narrow below due to self-pruning.
Densely contracted leaves (needles) are linear and exist in great number. They do not change with the seasons, but remain for 7–10 years. Tightly enveloping the stem, their blue-green color absorbs much light and creates a shadowy tree interior.	Extended on slender, flattened petioles, the shiny, roundish, finely-toothed leaves tremble and quiver with the slightest breeze, creating a dynamic play of light. They turn bright yellow/orange in autumn.
Develop best in 40–60% full shade and moist seed beds	Thrive on and require much sunlight (hence self-pruning of lower branches as shade increases).
Although very slow growing (1–5 inches tall after five years), they grow very old (up to 600 years) and very large (up to 150 ft with 6-foot trunk diameter). Lateral roots expand slowly (only 5 inches in first year), the weak taproot does not persist past juvenile years. Shallow root system makes it susceptible to wind throw.	Fast growing (5 ft. annually in early years), but shoots live only to an age of 70–80 years as individual trees (ramets) when light and open ground is available. (These are then replaced by shade-creating conifers.) Their rapidly growing lateral root system can sprout whole communities (up to 15,000/acre) of new ramets.
Monoecious, their seed cones become heavy and pendant, lignifying to become wooden. Scent does not dissipate, but withdraws into resins and oils.	Dioecious flowers produce catkins that dissolve into the atmosphere. Flower early, from March to May.
Dense seeds spiral downward close to tree and often penetrate soil.	Countless extremely light, tufted seeds (150,000–250,000/ounce) are carried miles by the wind before covering the forest floor like piles of cotton balls.
No seed production until age 15–40, large crops not produced until age 150–250.	Seed production already at ages 2–3, with large crops between ages 10–20.

Table I. Comparative Outline of Key Characteristics Found in Engelmann Spruce and Quaking Aspen

shimmering leaves, which creates a light-permeable environment that is welcoming and supportive of a diverse plant and animal community. Viewed from a somewhat larger perspective, one can observe, for example, that the striking contrast found in these two dominant tree species of the southwestern Colorado high country reflects the seasonal polarity of long, cold winters and short, sunlit summers. These seasonal alternations influence everything in that high altitude environment. When winter rules, low temperatures and long nights bring contraction and darkness. Life does not unfold under these conditions, but simply withdraws into a low-metabolism, hibernation-like state. On the other hand, the long days of summer are light-filled and warm. They call forth a celebration of colorful alpine plants that suddenly appear during that short season, only to withdraw again into their subterranean root systems when the winter darkness returns.

From this perspective it is possible to recognize qualities in these two species that express a fundamental gesture of the overriding ecosystem of which they are a part. Here, it appears to me, one can speak of the parts providing a reflection of the greater whole. This impression is strengthened when one considers this ecosystem from a temporal point of view. Just as summer follows winter, year after year, these two trees are cyclical in their relationship to one another and to their high country home. When the ancient, seemingly timeless spruce forests are disturbed through fire, avalanche, windfall, etc., quaking aspen unfold rapidly in their place. They pioneer the open ground and provide a summer-like habitat for several decades before the slow-growing Engelmann spruce gradually invade this space, creating shadows and needle-cover, crowding out the light-loving Aspen and leading, once more, to the dark, somber (winter-like) landscape of an Engelmann spruce forest.

In order to avoid an aspen-like sprouting into countless other possible comparisons, just one short mention of a different type of juxtaposition shall be included here. It is possible, for example, to highlight the contrast between our two tree species by viewing them in relation to the two formative (and quite polar) principles that are fundamental to all higher plants, namely, "stemming" and "leafing." The dominant form-tendency of the stem is narrow, compact linearity—linearity that in shrubs and trees lignifies to a seasons-transcending woody structure. As a chemically very inert substance, the lignin helps "seal off" this part of the plant from outer influences and gives it structural integrity. The typical leaf-blade, on the other hand, forms an expansive planar surface that opens itself to the light and air around it during the summer season, only to shrivel and die away with the approach of darkness and winter.

Viewed from this perspective, Engelmann spruce can be seen to possess a stem-like gesture from top to bottom. Not only is it spire-like in shape, linear in its leaves, and lignified in its pistillate cones, but it opens up very little to the influences of its surroundings and the seasons. Quaking aspen, on the other hand, is very leaf-like in its openness to the light (it photosynthesizes right down to its primary stem, the trunk), in its trembling, air-movement-sensitive leaves. Quaking aspen reacts rapidly to changes in its environment as it expands into disturbed areas and is quick to grow in the short summer season available to it. With its tiny seeds it gives itself over to the air and the wind, the radiant color of its autumn leaves rays back some of the light absorbed during the short summer months.

In the contrast between these two plants we see expressed what Goethe found so fundamental to the unfolding of the plant: the gestures of expansion and contraction. Whereas the plant is linear and contracted in its inwardly oriented conductive system, it is expansive and peripherally open in the sun , air and light-connected surfaces of its leaves. It is not difficult to see in both Engelmann spruce and quaking aspen how a particular tendency is one-sidedly emphasized.

Pedagogical Perspectives

One key benefit of the comparative method is how it awakens a sense for gesture when one moves back and forth between the things being compared. For example, if you move from a spruce leaf (needle) to an aspen leaf in active observation—you sense a gesture of stretching out into the periphery through the needle-like petiole and the a tremendous horizontal expansion and thinning out of substance as you move to the leaf blade—what was closed and tight before, is now extended outward and totally opened up as it interfaces (literally) with what is around it. This experience can be intensified and brought to greater awareness by creating a mental picture the first object (rather than observing it directly) and then transforming that image (and its various qualities: density, color, surface texture etc.) into the second one. Because we construct the first image in our imagination without the support of the finished sense perception, we experience how our inner image-building activity must change when moving from one form to the next. We experience the formative gesture that we must produce in order to bring about the transformation—we are becoming aware of our own productive activity as we "recreate in the wake of ever-creating nature." (Goethe) What is being compared plays a huge role in determining which qualities move to the foreground of our experience, but central in all cases is that we move from a static, finished experience of an object to a dynamic one that awakens the sense of something "coming into being"—of formative activity.

In the case of the two trees discussed above, moving in our imagination from a spruce needle to the expansive, quivering leaf blade of an aspen brings a decisive experience of opening up to, of interacting with the surrounding light and air. Going from aspen leaf to spruce needle, on the other hand, creates the opposite gesture of shrinkage, densification, withdrawal into itself....

If we transform the overall form, color, and mood of an Engelmann Spruce into a Quaking Aspen (or a stand of Engelmann Spruce into a stand of Quaking Aspen), we have a similar experience: one of opening up, of engaging the periphery. We can, in fact, move down through Table I in this way, transforming the left column into the right one, and repeatedly experience a similar gesture. (Or move from the right column to the left and experience the opposite gesture.)

This comparative/transformative approach helps us to sense forms and qualities in nature more dynamically. It is helps us move from what Goethe called the *Gestalt* to another level of experience that he referred to as *Bildung*:

> *The Germans have a word for the complex of existence presented by a physical organism:* Gestalt *(structured form). With this expression they exclude what is changeable and assume that an interrelated whole is identified, defined, and fixed in character. But if we look at all these Gestalten, especially the organic ones, we will discover that nothing in them is permanent, nothing is at rest or defined—everything is in a flux of continual motion. This is why German frequently and fittingly makes use of the word* Bildung *(formation) to describe the end product and what is in process of production as well. Thus, in setting forth a morphology we should not speak of* Gestalt, *or if we use the term we should at least do so only in reference to the idea, the concept, or to an empirical element held fast for a mere moment of time. When something has acquired a form, it metamorphoses immediately to a new one. If we wish to arrive at some living perception of Nature, we ourselves must remain as quick and flexible as Nature and follow the example she gives.* (Goethe, in Miller 1988, pp. 63–64)

One of the primary goals of Waldorf education is to bring mobile, living concepts to students that are able to unfold organically within them as they grow older. In this essay I have attempted to discuss and illustrate why the comparative method can be an important aid in realizing that goal. In contrast to static definitions that "fix or mark the limits of" a few identifying traits unique to what is being considered, the comparative (transformative) approach steps across those isolating boundaries and helps awaken process-related experiences that—instead of fixing (*Gestalt*)—reveal gesture and formative tendencies (*Bildung*).

Comparisons are open-ended, they are "conversations" between the phenomena being compared, and each conversation is unique to those particular conversing partners. When compared to an oak, Quaking Aspen will "speak" somewhat differently to us than when it is contrasted with a beech, or a rose bush. Engelmann Spruce will "speak" in more subtle tones when transformed (in our imagination) into a subalpine fir rather than a Quaking Aspen. In this essay, our choice of the organisms to be compared was determined by their striking dominance in the area of high-country Colorado where they were experienced. Through these concluding remarks, it should be evident, however, that the possibilities for further comparative conversations between the two trees we have focused upon and the many other organisms found the same region are great. In fact, these possibilities are only limited by the time and energy we are willing to invest in providing the opportunity for various organisms to speak of themselves and each other upon the inner stage of our comparison-capable consciousness.

Resources

Alexander, R. and Shepperd, W. (1990). "Engelmann Spruce," In Burns, R. and Hankala ,B., eds. *Silvics of North America.* Washington DC: USDA, Forest Service.

Braun, E. (2001). *Deciduous Forests of Eastern North America.* Caldwell, NJ: Blackburn Press.

Farrar, J. (1995). *Trees of the Northern United States and Canada.* Ames: Iowa State University Press.

Goethe, J. (1988). *Goethe: Scientific Studies.* Miller, D, ed. New York: Suhrkamp.

Grant, M. and Mitton, J. (2010). *The Glorious, Golden, and Gigantic Quaking Aspen.* Nature Education Knowledge 1(8): 40–44.

Grohmann, G. (1996). *The Plant.* Kimberton, PA: Biodynamic farming and Gardening Assoc.

Holdrege, M. (2009). *From Beach to Savanna: Vegetation Transitions at Illinois Beach State Park*. Ann Arbor: ProQuest LLC.

Holmes, B. (1999). *Quaking Aspen*. Minneapolis: Carolrhoda Books.

Johnson, D. (2009). *Biogeography of Quaking Aspen [Populus tremuloides]*. http://bss.sfsu.edu/geog.

Julius, F. (1969). *Metamorphose*. Stuttgart: JC Mellinger.

Kharouk, F., Middleton, E., et al. (1995). "Aspen Bark Photosynthesis and Its Significance." *Later, Air and Soil Pollution* (82): 483–497. Netherlands: Kluwer Academic Publ.

Margulis, L. and Sagan, D. (2000). *What Is Life*. Berkeley: University of California Press.

Miller, D. (1988). *Goethe—Scientific Studies*. New York: Suhrkamp.

Mitton, J. and Grant, M. (1980). "Observations on the Ecology and Evolution of Quaking Aspen, Populus tremuloides, in the Colorado Front Range." *American Journal of Botany* 67(2): 202–209.

Naydler, J. (1996). *Goethe on Science*. Trowbridge, GB: Cromwell Press.

Perala, D. (1990). "Quaking Aspen." In Burns R and Hankala B, eds. *Silvics of North America*. Washington DC: USDA, Forest Service.

Powell, G. (2009). *Lives of Conifers*. Baltimore: Johns Hopkins University Press.

Preston, R. and Graham, R. (2002. *North American Trees*. Ames: Iowa State University Press.

Steiner, R. (1990). *The Study of Man* (CW 293). London: Rudolf Steiner Press.

_______ (2002). *The Light Course* (CW320). London: Rudolf Steiner Press.

Tudge, C. (2005). *The Tree*. New York: Three Rivers Press.

APPENDIX III.
Studying a Local Ecosystem with Students—an Example

In what follows, an example is given of what the on-sight investigation of a specific ecosystem could look like in the 11th grade. The ecosystem in question, Illinois Beach State Park (IBSP), lies on the eastern shore of Lake Michigan, just south of the Wisconsin/Illinois border.

Non-phylogenetic Plant Classification

One of the important decisions that a teacher must constantly make—one that has been frequently referred to in this book—is which details are important to focus on and which are not, so that the students do not lose sight of the forest through the trees. In the case of ecosystems, this metaphor can be taken literally: It is easy to get so caught up in identifying and naming the plants so that we lose track of the overall ecosystem-characteristics that we are hoping to discover.

A helpful way to avoid an overemphasis on plant identification, while still attending to how the plant population is changing, is to focus on non-phylogenetic plant classifications rather than the classical Linnean taxonomy. In recent decades, increasing interest in non-phylogenic plant classifications has been awakened by the challenges of predicting the effects of climate change and agricultural land use on terrestrial ecosystems. A way has been sought that simplifies the vast diversity of nature into something that is more manageable and that can bridge the gap between specific plant species and ecosystem processes. Determining which set of traits best indicates environmental change has been a topic of much debate and has led to the use of many different approaches in the field. (Diaz & Cabido 1997; Gitay & Noble 1997)

In this study I use the term "growth-form" to refer to morphological features of the plants under consideration. As far as the choice of growth-form categories is concerned, I share the perspective of one of the pioneers of growth-form studies, C.H. Gimmingham, who concluded that the best results are likely to be achieved by using growth-form classifications "developed independently for each study, based on characters which are observed in the field to be of importance in the communities examined" (1951, p 398).

Phenotypic Plasticity in Relation to Plant Available Light (PAL), Plant Available Moisture (PAM), and Plant Available Nutrients (PAN)

One effective growth-form perspective for assessing the changes in vegetation as one moves from the shoreline landward at ISBP is to pay attention to how the plant's morphological tendencies change in relation to the amount of moisture, light and soil nutrients available to them. It is well known, of course, that changes in plant available moisture (PAM) and plant available light (PAL) have contrasting effects on plant life. As most of the students will have learned, it is common practice among gardeners to give plants plenty of water and humus-rich soil when enhanced leaf-volume is desired, and less of these to plants when the flowering component is to be emphasized (Julius 1969). Field experiments confirm the validity of such practices. Plants exposed to higher levels of light and warmth have been found to develop shorter petioles and more pointed and differentiated leaves than the rounder, softer, longer-petioled leaves found on plants that receive higher levels of water and more fertile soil (Schaumann 1982).

The capacity of plants such as the dandelion (*Taraxacum officinale*) to produce different phenotypes in response to distinct environmental conditions (phenotypic plasticity) also shows this quite vividly. When grown in a shadowy location on rich soil, dandelions will develop tall stems and large leaves that show little indentation, whereas those grown in direct sunlight and with less plant available nutrients (PAN) have short stems and rosettes of small, finely chiseled, linear leaves (Julius 1969; Pigliucci 2001). When grown in shade, the thistle (*Cecerbita muralis*), develops uniformly green, long-petioled and large-surfaced leaves. Grown in full light its leaves are yellow in tone and their edges reddish, the petioles are less developed, and the more highly indented and segmented leaf blades create less surface area. Even the common radish (*Raphanus sativus*) develops smaller, denser and more formed leaves (less peripheral expansion) when grown with higher levels of light than it does under shady conditions, when the leaves expand to a larger, dark-green leaf-blade that extends outward on a long petiole. (Bockemühl 1981)

To sum up: An increase in PAM leads to greater vertical growth and also expansion laterally through branching, petiole development, and expansive widening of the leaf blade surface. As the counter pole to that, increased light intensity (in the absence of increased PAM) leads to contraction in the plant: less vertical growth, less branching higher on the central stem, fewer and shorter petioles, and narrower leaf-blades.

With these morphological distinctions in mind, the students can observe the way in which the vegetation changes at IBSP in regard to verticality (central stem development) and horizontality (petiole length, expansiveness of the leaf blade, and stem branching) as we move inland from the shoreline.

The Setting

As noted above, IBSP is located in northern Illinois on the western shore of Lake Michigan. It encompasses more relatively undisturbed and high-quality natural communities than anywhere else in the State of Illinois, and consists of a series of linear sand and/or gravel ridges that run parallel to the coastline (Greenberg 2002).

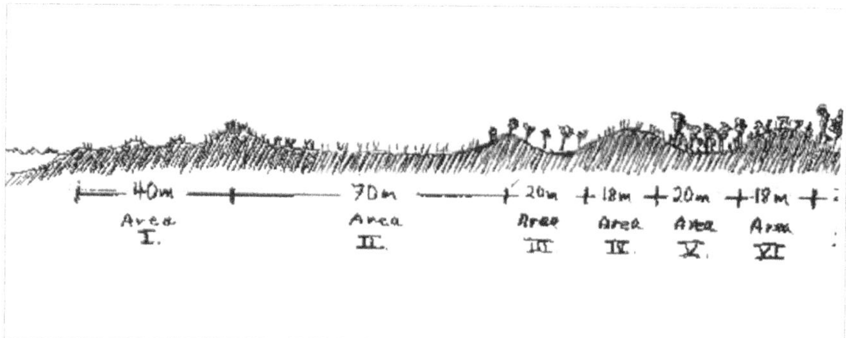

Fig. III-1 Profile of terrain at Illinois Beach State Park (IBSP) – east to west.

This washboard topography of alternating ridges and swales is about 1.7 km wide in the area of IBSP. The elevation difference between ridges and swales is normally no more than 3.5 m. Starting at the water's edge, the first task is to observe and discuss with the students onsite the environmental factors just mentioned—Plant Available Light (PAL), Plant Available Moisture (PAM), and Plant Available Nutrients (PAN)—as well as the other relevant variables such as wind intensity, climate, possible pollutants, etc.

With the help of occasional input from the teacher, the initial findings at IBSP would be something like the following. The surface sediments consist of concentrated coarse sand, pebbles, and cobbles; whereas further inland the dunes are formed by wind action that blows fine sand from dry sections of the beach into the back beach area. Dune communities experience high levels of light exposure. Moreover, the intensity of direct illumination is increased greatly by reflection: On a bright summer day, the glare of the white sand is extreme. Wind is a potent factor in shaping life along the dunes. Onshore winds frequently pick up sand and carry it farther inland. On windy days, a natural sandblast-effect is evident. The wind

Fig. III-2 The lakeshore—looking eastward.

can also have a destructive effect on plant life by blowing the soil beneath them away—thus leaving the roots exposed above the surface—or by advancing a dune that buries the plants in its path.

The soil of the dunes is mostly disintegration-resistant quartz. The light color of quartz particles is a major factor in the high light-reflectivity of the sand. Dune sand is remarkably uniform in size when compared with beach sand, because the wind is unable to transport beach gravel or larger sand particles for any distance. Because of the relatively large size of its constituent particles, beach sand is extremely porous and almost without cohesion between grains. This is significant in regard to heat relations, as mentioned above, but also results in the sand's weak capacity to hold water: In sand, water percolates rapidly downward into the groundwater below. As a result, the low levels of PAM result in conditions reminiscent of the desert.

The surface sediment of concentrated, coarse sand, pebbles and cobbles gives way to windblown medium grain sands as one moves landward. Not unlike the desert, temperature variances are more extreme than elsewhere in the park. High levels of light exposure (PAL)—the intensity of which is greatly increased by reflection—combines with the low specific heat of the sand to let the surface temperature rise to over 100 degrees F in the summer and plummet below zero in the winter.

Fig. III-3 The lower beach.

Changing Growth Forms

The Beach

When one examines the seven plant species found in the beach area of IBSP, it is striking that none of their leaves have petioles. This means that the leaves hug the stem, that there is very little expansion laterally out into the surrounding light and air. This hugging tendency is strongly accentuated in the three grass species found here—dune wheat grass (*Agropyron daststachyum*), marram grass (*Ammophila breviligulata*) and sand reed (*Calamovilfa longifolia*)—in that the lower portion of their leaves literally enclose the stem as so-called leaf-sheathes before finally bending outward and away from the stem as narrow, linear leaf-blades.

Fig. III-4 Dune wheat grass [*Agropyron dasystachyum*].

Observing the leaves of the non-grasses in this area, one finds that those of *Salsola tragus* (Russian thistle) are sharp-prickly while the leaves of beach wormwood (*Artemesia caudata*) are narrow and finely dissected into linear segments. The tiny leaves of seaside spurge (*Euphorbia polygonifolia*), in contrast, contain within them a milky, water-preserving sap, and those of sea rocket (*Cakile edentula*)—though small—are thick and fleshy (succulent). The sea rocket possesses a horizontal stem that sends up a few sparse shoots and leaves in some locations, while in others it may grow somewhat bushy.

Fig. III-5 Sea Rocket. [*Cakile edentula*]

The stems of all three grasses are primarily prostrate and underground. From their horizontal buried stems (rhizomes) they send up vertical shoots that appear above ground with the narrow linearity described above in reference to their leaves. This horizontal (rather than vertical) extension of the central stem is also shared by two of the four dicot species. In short, the plants of the beach show very little of the vertical and peripheral unfolding that typify the stems and leaves of most plants. In many regards, the beach reminds one strongly of the kinds of conditions that one finds in the desert regions of the earth. Interestingly, this similarity does not stem from a lack of moisture, as such, as is the case in the ITCZ-influenced deserts, but from a lack of plant *available* moisture (PAM), resulting from the porosity of the sand, as noted above, and a light-intensity that is not only comes from a lack of shade, but also results from the high reflectivity of the quartz-rich beach sand.

Fig. III-6 Seaward side and crest of first beach ridge. (This ridge builds the western boundary of the beach area.)

Fig. III-7 Leeward side of beach ridge III-6.

Beach Ridge and Swale One

The contrast between the seaward and the leeward side of the sand ridge that marks the boundary between the beach and the first swale is striking. Whereas the percent of sand cover on the beach is about 90–95 percent, it drops to 50–60 percent in the swale. In the wind shadow of the sand ridge, the force of the wind decreases greatly. The leeward side of the beach ridge runs down into the wind-shadow of the swale. Shrubby patches are found in this wind-shadow. They spread by underground runners (rhizomes) and form thicket-clusters. On normal summer days, the sea breeze does not reach this area and the ambient heat increases noticeably compared with the breeze-cooled beach air.

Moving into the flatter terrain at the eastward edge of the swale (60% sand cover), a very different quality becomes apparent in the vegetation, one that dominates the rest of this 70-meter-wide swale. Visually, one is struck by the verticality of the flora here: Tall erect stems and sessile, upwardly inclined, narrow leaves prevail. Scattered here and there are rounded clumps formed by sand cherry and blue-leaved willow shrubs. These make, however, only a secondary impression next to the striking verticality of the other plants. The overall color-tone is purplish.

Fig. III-8 First swale with flowering Rough Blazing Star [*Liatris aspera*].

The impression of verticality is largely created by the presence of a contingent of seven forb species. Leading the way is the Rough Blazing Star (*Liatris aspera*), with its narrow, lanceolate, sessile leaves. Interspersed among the forbs is the conspicuous presence of little bluestem, and a scattering of sand reed, both grasses. All in all, it is the strong vertical growth with very little or no lateral branching that is striking.

One has the impression that the vertical central stem has moved firmly into the foreground in this section of IBSP. In the case of the forbs, This narrow upright impression is accentuated by the fact that most of the leaves show little horizontal expansion—their narrow surfaces not only hug the stem in most cases, but many of them tend to tilt upward rather than outward. In comparison to the beach, this area of IBSP still shows little peripheral leaf development, but a very definite step "upward" in regard to the vertical extension of the stem. Tall and narrow dominates in the swale: In the vertical dimension the plants are beginning to expand, peripherally they remain contracted.

The shrubs, on the other hand, emphasize a different aspect of the stem. Their lignin-permeated stems bring to the fore lateral extension, structural permanence, and autonomy with respect to the outer environment. This unfolding is significant when compared to the beach area to the east, but modest in relation to what develops as one moves farther inland. The leaf aspect, on the other hand, has yet to expand significantly, they retain a narrow linearity that is even more accentuated than that found on the beach.

Swale Two and Ridge Two

The second swale is an area approximately 20 meters across and includes, as its eastern flank, the leeward side of the second ridge. Moving westward from the lake, one finds here for the first time an increase in PAM (from 0.0 to 0.8) and a decrease in PAL (from 6.5 to 3.3). The soil, although clearly still sand, is significantly darker than what we have seen up until now.

Fig. III-9 Swale two.

This area also makes a very different impression on the observer: Large leaf-surfaces—the likes of which have been completely absent up to this point—catch one's attention immediately (Fig. III-9). The foliage is also much denser and the dominant color-tone changes from "purplish" in Swale One to a satiated greenish-blue here. This area has a much moister, "juicier" feel to it, and for the first time it is no longer possible to see the sand beneath one's feet—it is now covered with a thick layer of leaf and stem litter. The presence of large patches of shade provides a conspicuous contrast to the direct sunshine that has prevailed up to this point. The shade comes from five black oaks (*Quercus velutina*) that range in height from three to seven meters. A few shrubs also make their home here.

The strong greenish-blue, broad-leaf impression that dominates the scene emanates from yet another plant, however. It comes from the woody vine (liana) *Vitis riparia* (riverbank grape). This plant covers roughly one fourth of the surface near ground level, before it begins to climb upward into the oaks and the tallish shrubs.

As far as the stems are concerned, an interesting polarity is noticeable. On the one hand, the vertical extension of the plant, represented by trees, comes fully into its own for the first time. The central stem (axis) of these plants—the trunk of the black oaks—is now so highly developed that only a few specimens are able to create sufficient foliage to shade roughly 70% of this area from a height of up to seven meters.

At the other extreme, we find the woody vines. These lack entirely the principal characteristic of the central stem as it has been discussed up to this point: vertically. Instead, the lateral extension of the petiole and the widening of the leaf-blade dominate in these plants. Because of this one-sidedness—their weak internodes grow outward rather than upward—these vines must depend on the stems and branches of other plants in order to wind their way up into heights that they cannot reach on their own. This tendency to petiole development and wide leaf blades is not limited to the vines in this swale. It has now become the rule rather than the

Fig. III-10 Ridge three: Looking southward along the dome of the ridge.

exception. For the first time, the expansion of the plant into both the vertical and the horizontal begins to take hold at IBSP.

Ridge Three

The third ridge landward at IBSP rises roughly 2m above the concavity of Swale Two. Compared to Ridges One and Two—that have little or no crown—this ridge is much wider (Fig. III-10). It forms a dome-like arch that stretches between the swales to the east and west of it. Conditions on this ridge are quite different from Swale Two. PAM falls from 0.8 in the swale to 0.0 on the ridge, while PAL increases from 3.3 to 6.5. The surface soil, however, remains similar to that in Swale Two: dark brown sand. With this drop in moisture-availability and rise in light penetration, it is not surprising that the leaves grow narrower again, while the median petiole length drops. Along with this, the strong vertical expansion found in the trunks of the oak trees located in Swale Two vanishes almost completely. It is now present only in the form of several small black oaks that can be found scattered about on the ridge-dome. Since the decrease in PAM and increase in PAL creates conditions similar to those found on the beach and in the first swale, it is not surprising that the vertical and peripheral unfolding that developed in Swale Two, contracts once more on Ridge Three and returns to levels similar to those found on the beach and in Swale One.

Fig. III-11 Looking westward from the dome of ridge three down into swale three.

Swale Three

The third swale shows a drop in PAL to 3.3 compared to the 6.5 reading for Ridge Three. On the other hand, the less elevated, run-off-receiving swale shows an increase in PAM to 1.2 (from ~0.0 on the ridge above). It is thus not surprising, as Fig. III-11. makes evident, that the descent from the dome of Ridge Three into Swale Three shows a significant increase in the density and dimensions of its vegetation. Swale Three is very reminiscent of Swale Two, except that it advances the vertical stem tendency even further with a set of black oaks that reach a height of 15 meters or more. The shrubs and vines that were completely absent on Ridge Three unfold here in a similar manner to Swale Two. Put differently, one finds not only an increase in verticality, but also in horizontal expansion as one moves down into this damper and darker swale from the drier and more light-exposed ridge that went before.

Ridge Four

On ridge four the pattern developing in the preceding plots repeats itself. Corresponding to the previous convex ridge, with a decrease in PAM and an increase in PAL comes contraction: All three categories register a drop (height decrease, fewer petioles, narrower leaf blades) relative to the preceding swale.

Fig. III-12 Ridge four.

Fig. III-13 Swale four.

Swale Four

Swale four confirms the pattern. This descent into yet another swale leads to increased PAM, decreased PAL, and plant expansion. Here, at the highest recorded levels of PAM, the greatest vertical and peripheral expansion are evident.

Such high correlation between increasing PAM and decreasing PAL is not surprising when one considers their concrete relationship at IBSP. Decreases in PAL result from patches of shade created by the increased growth of shrubs and trees. The latter has a double effect on PAM. First, such growth generates additional ground litter that absorbs moisture that otherwise would percolate rapidly downward in the sandy soil at IBSP. Second, the shade reduces moisture evaporation caused by direct sunlight. These effects lead to an increase in PAM that, in turn, leads to an increase in the growth of shrubs and trees, which decreases PAL, and so on.

Overview

If, stepping back, we now compare the changes in vegetation from the beach all the way to Swale Four, we see a significant transformation.

On the one hand, we have been able to observe many examples of micro-habitats. These were based on changes in PAL, PAM and PAN that evolved as we moved inland from the lakeshore. On the other hand, we have witnessed successive changes in the vegetation that—when they were discovered for the first time—played a pivotal role in the development of a new scientific discipline.

Fig. III-14 On the beach.

Fig. III-15 Swale four.

Fig. III-16 And finally, further inland, we find ourselves in a diverse oak savanna.

Ecological Succession

What the students have been able to observe at IBSP is related to phenomena discovered not far down the shoreline at the southern end of Lake Michigan. What was observed there led to the formulation of the central ecological principle known as "succession" (mentioned earlier in this book), which would become one of the central ideas in the unfolding of modern ecology. The story—in short—goes like this:

In the 1890s, as a graduate student and later professor at the University of Chicago, Henry Cowles began traveling to the southern shore of Lake Michigan. Moving from water's edge over a series of rolling dunes, Cowles noticed a striking sequence of changing vegetation as he moved from the youngest lakeside dunes to the older dunes further inland, which culminated in a mixed oak forest with a thick layer of humus. Cowles realized that walking inland from the edge of Lake Michigan he could trace a *spatial* pattern of ecological succession that was similar to how that vegetation must have developed over *time*. In the rapidly changing dune landscape, he could see spread out in space—in the present—a process that had taken place sequentially over long periods of time in the past. His observations opened up whole new ways at looking at nature and were a key factor in the establishment of ecology as a science.

If time allows, the students benefit greatly from taking a field trip to the Indiana Dunes (Miller Woods), where Cowles made his discoveries.

Resources

Allaby, M. (1998). *Oxford Dictionary of Ecology*. 2nd ed. Oxford: Oxford University Press.

Bockemühl, J. (1981). *In Partnership with Nature*. Wyoming, RI: Bio-Dynamic Literature.

Chapin, F., Matson, P., and Mooney H. (2002). *Principles of Terrestrial Ecosystem Ecology*. New York: Springer.

Cowles, H. (1899). "The Ecological Relationships of Vegetation on the Sand Dunes of Lake Michigan." *Botanical Gazette* 27: 95–117. Also reprinted in Real, L. and Brown, J., eds. (1991). *Foundations of Ecology, Classic Papers with Commentaries*. Chicago: University of Chicago Press.

Diaz, S. and Cabido, M. (1997). "Plant Functional Types and Ecosystem Function in Relation to Global Change." *Journal of Vegetation Science* 8: 463–474.

Gimmingham, C. (1951). "The Use of Life Form and Growth Form in the Analysis of Community Structure as Illustrated by a Comparison of Two Dune Communities." *Journal of Ecology* 39(2): 396–406.

Gitay, H. and Noble, R. (1997). "What Are Functional Types and How Should We Seek Them?" In Smith T, Shugart H, and Woodward F. *Plant Functional Types*. Cambridge: Cambridge University Press, 3–19.

Greenberg, J. (2002). *A Natural History of the Chicago Region*. Chicago: University of Chicago Press.

Holdrege, M. (2009). *From Beach to Savanna: Vegetation Transitions at Illinois Beach State Park*. Ann Arbor: ProQuest LLC.

Imes, R. (1990). *The Practical Botanist*. New York: Simon & Schuster.

Julius, F. (1969). *Metamorphose*. Stuttgart: JC Mellinger.

Pigliucci, M. (2001). *Phenotypic Plasticity*. Johns Hopkins University Press

Rodd, T. and Stackhouse, J. (2008). *Trees: A Visual Guide*. West Sussex: Bonnier Books.

Schaumann, W. (1982). "Die Bildung der Pflanzenqualität als Ergebnis der Wirkungen von Erde und Sonne." In Schad, W., ed. *Botanik*. Stuttgart: Verlag Freies Geistesleben.

Index

Acknowledgments

My sincere thanks go out to those individuals—E.M. Kranich, W. Schad, G. Zickwolff, and J. Bockemühl—who first awakened in me many of the questions addressed in this book. More recently, I wish to acknowledge what a significant resource the work of Craig Holdrege and Steve Talbott at the Nature Institute has been in exploring many of the topics discussed here. (Thanks to Craig, too, for "outing" all those sentences in the book whose tense might be considered "passive" in nature!). Mark Riegner also deserves a word of thanks for helping guide me through my first explorations into the realms of ecology.

At Waldorf Publications, the enthusiasm and support that Patrice Maynard has brought to this project has been very enlivening. The work that she, Ann Erwin, and Mary Stowe put into helping such a volume become an earthly reality is immense and can't be thanked enough.

My wife Patricia also deserves a word of gratitude. The love and enthusiasm that she brings to the natural world around her is truly inspiring—she walks the talk!

Lastly, but actually first and foremost, I want to express my deep gratitude for the role that Rudolf Steiner's life work has played in all my endeavors. He created, in innumerable ways, deep foundations for finding "windows into the dynamic nature of living organisms."

Made in the USA
Monee, IL
12 September 2024

65261446R00181